CHEMISTRY RESEARCH AND APPLICATIONS

AN INTRODUCTION TO MELAMINE

CHEMISTRY RESEARCH AND APPLICATIONS

Additional books and e-books in this series can be found on Nova's website under the Series tab.

CHEMISTRY RESEARCH AND APPLICATIONS

AN INTRODUCTION TO MELAMINE

ASHLEY HARRIS
EDITOR

Copyright © 2020 by Nova Science Publishers, Inc.

All rights reserved. No part of this book may be reproduced, stored in a retrieval system or transmitted in any form or by any means: electronic, electrostatic, magnetic, tape, mechanical photocopying, recording or otherwise without the written permission of the Publisher.

We have partnered with Copyright Clearance Center to make it easy for you to obtain permissions to reuse content from this publication. Simply navigate to this publication's page on Nova's website and locate the "Get Permission" button below the title description. This button is linked directly to the title's permission page on copyright.com. Alternatively, you can visit copyright.com and search by title, ISBN, or ISSN.

For further questions about using the service on copyright.com, please contact:
Copyright Clearance Center
Phone: +1-(978) 750-8400   Fax: +1-(978) 750-4470   E-mail: info@copyright.com.

NOTICE TO THE READER

The Publisher has taken reasonable care in the preparation of this book, but makes no expressed or implied warranty of any kind and assumes no responsibility for any errors or omissions. No liability is assumed for incidental or consequential damages in connection with or arising out of information contained in this book. The Publisher shall not be liable for any special, consequential, or exemplary damages resulting, in whole or in part, from the readers' use of, or reliance upon, this material. Any parts of this book based on government reports are so indicated and copyright is claimed for those parts to the extent applicable to compilations of such works.

Independent verification should be sought for any data, advice or recommendations contained in this book. In addition, no responsibility is assumed by the Publisher for any injury and/or damage to persons or property arising from any methods, products, instructions, ideas or otherwise contained in this publication.

This publication is designed to provide accurate and authoritative information with regard to the subject matter covered herein. It is sold with the clear understanding that the Publisher is not engaged in rendering legal or any other professional services. If legal or any other expert assistance is required, the services of a competent person should be sought. FROM A DECLARATION OF PARTICIPANTS JOINTLY ADOPTED BY A COMMITTEE OF THE AMERICAN BAR ASSOCIATION AND A COMMITTEE OF PUBLISHERS.

Additional color graphics may be available in the e-book version of this book.

Library of Congress Cataloging-in-Publication Data

Names: Harris, Ashley (Nova editor), editor.
Title: An introduction to melamine / Ashley Harris, editor.
Description: New York : Nova Science Publishers, 2020. | Series: Chemistry
   research and applications | Includes bibliographical references and
   index. |
Identifiers: LCCN 2019059676 (print) | LCCN 2019059677 (ebook) | ISBN
   9781536171365 (paperback) | ISBN 9781536171686 (adobe pdf)
Subjects: LCSH: Melamine. | Triazines--Toxicology. | Food--Toxicology.
Classification: LCC QD401 .I58 2020 (print) | LCC QD401 (ebook) | DDC
   615.9/5142--dc23
LC record available at https://lccn.loc.gov/2019059676
LC ebook record available at https://lccn.loc.gov/2019059677

Published by Nova Science Publishers, Inc. † New York

Contents

Preface		vii
Chapter 1	Melamine and Its Analogous: Analytical Methods and Related Levels in Foodstuffs *Mena Ritota and Pamela Manzi*	1
Chapter 2	Influence onto the Immatures: Melamine from Mothers to Babies *Ching Yan Chu and Chi Chiu Wang*	39
Chapter 3	Use of Melamine-Formaldehyde Resin as Shell Material for Microencapsulation *María de la Paz Miguel*	77
Chapter 4	Melamine Sensor Development Based on Mixed Metal Oxide Nanoparticles *Mohammed M. Rahman, Abdullah M. Asiri and M. M. Alam*	115
Chapter 5	Formation of Melamine-Derived Particles in Aqueous and Biological Matrices *N. S. Chong, D. Dutta and B. G. Ooi*	141

| **Chapter 6** | Melamine Detection with Nanostructure Materials
Mohammed Muzibur Rahman | **169** |

Index   **185**

Related Nova Publications   **189**

PREFACE

In this compilation, technical and critical aspects concerning confirmatory methods for the analysis of melamine and its analogous, including cyanuric acid, ammeline, and ammelide in foodstuffs are discussed. Moreover, an overview of the concentrations of melamine and related compounds reported in the recent literature for different food items are summarized, as the results from monitoring these compounds in food products indicate their continuing presence in the food chain.

Although high level of adulteration has ceased, melamine is still a popular material for pesticides, farm animal feed fillers, fire retardants, anti-wrinkles and mild abrasives. As such, the authors discuss how low-dose contamination of melamine to the environment cannot be ignored.

Next, melamine-formaldehyde resin was chosen to form shell material due to its good thermo-mechanical and water-resistant properties. A series of experiments are conducted on changing the emulsifying system and a discussion of the results is provided.

Additionally, a sensitive chemical sensor is developed to detect melamine selectively by an electrochemical approach, where ternary mixed metal oxide nanoparticles were prepared through the wet-chemical process. The calcined $ZnO/CuO/Co_3O_4$ NPs are investigated by field emission scanning electron microscopy, energy-dispersive X-ray spectroscopy, X-ray photoelectron spectroscopy, powder X-ray diffraction, ultraviolet visible spectroscopy, and Fourier-transform infrared spectroscopy.

The authors evaluate the formation of melamine-derived particles in aqueous and biological matrices using different analytical techniques for studying the bioaccumulation of melamine-cyanurate in tissues, including kidney stones.

Lastly, cadmium doped antimony oxide nanostructures are synthesized by a facile wet-chemical method at a low temperature to detect melamine from aqueous solutions. The calcined cadmium doped antimony oxide nanostructures are characterized systematically by FE-SEM, EDS, UV/Vis., FTIR spectroscopy, powder XRD and XPS techniques.

Chapter 1 - In the past decade, adulteration incidents concerning the fraudulent addition of melamine in dairy products and vegetable proteins occurred in China prompted numerous governmental and private laboratories to develop analytical methods for melamine and related compounds determination in foods. Since then, many efforts have been made to develop both screening and confirmatory methods based on traditional analytical techniques (liquid and gas chromatography, mass spectrometry, liquid and gas chromatographies coupled to mass spectrometry, capillary electrophoresis) and increasingly innovative techniques (i.e., immunoassay, sensor technology, vibrational spectroscopy).

In this chapter, technical and critical aspects concerning confirmatory methods for the analysis of melamine and its analogous, including cyanuric acid, ammeline, and ammelide, in foodstuffs will be discussed. Moreover, an overview of the concentrations of melamine and related compounds reported in the recent literature for different food items (i.e., dairy products, meat, seafood, egg, vegetables, soy and cereal products) will be summarized, as the results from monitoring these compounds in food products indicate their continuing presence in the food chain. Sources of baseline melamine and cyanuric acid in foodstuffs, in fact, may come from melamine-containing plastic materials for tableware, herbicide/pesticide and fertilizers use, or from the industrial production of melamine resins, which may be an ongoing source of melamine in water and the environment in general. At the same time, melamine can be degraded *via* deamination reactions to its analogues ammeline, ammelide, and cyanuric acid.

Melamine and cyanuric acid are the major compounds observed in foods, ammeline and ammelide generally accounting for less than 10% of the total content. Fortunately, current melamine levels reported in foods are well below the tolerance limits set by Food and Drug Administration (2.5 mg/kg and 1.0 mg/kg in milk products and powdered infant formula, respectively).

Chapter 2 - In the last 2 decades, melamine contamination of animal feed and infant formulae has raised a global concern on food safety. Since then researches on the compound manufactured for the past 50 years have been re-activated. Most of the studies mainly focused on renal toxicity. Timely clinical treatments to the affected babies of the scandal have been convincingly beneficial. However, studies of reproductive and developmental toxicity of melamine on pregnant mothers, developing foetuses and neonates have been lacking. Although high level of adulteration has ceased, melamine is still a popular material for pesticides, farm animal feed fillers, fire retardants, anti-wrinkles and mild abrasives, etc. Low-dose contamination of melamine to the environment cannot be ignored. Animal models demonstrated the low-dose melamine transfer from mothers to foetuses and/or neonates, toxicokinetics in the pregnant, foetal and neonatal rats and reproductive and developmental toxicology at different stages of development. Detailed studies on the long-term effects on the immatures are still very limited.

Although there have been no clinical reports of this kind of data, animal experiments have proven the transfer of melamine from the mothers to their offspring, with evidence in birds and mammals in particular. Presence of melamine in eggs, muscles and various organs of the offspring has been reported. Loss of embryos, delay of foetal development, morphological changes in the kidneys and bone forming centres have also been revealed. On the other hand, it is also worth-considering the change of the maternal physiology during pregnancy that ingestion of melamine in daily diet or the contaminated maternal formulae can post risks that are different from those to the non-pregnant individuals. As it is well known that kidneys can compromise during pregnancy, an extra stress exerted by melamine may further complicate the condition. With these evidences, the authors can postulate that acute kidney injury may not be the whole story of melamine effects on the infants.

Chapter 3 - Polymers are commonly used in the fabrication of protective coatings for metallic substrates. Research on self-repairing coatings based on the incorporation of microcapsules or nanocapsules loaded with corrosion inhibitors or film-former healing agents continues to increase because of potential economic benefits of this technology. This chapter deals with the use of melamine-formaldehyde resins as shell material for microcapsules. The microencapsulation of linseed oil as repairing agent was carried out via the in-situ emulsion polymerization method. Chemical composition of microcapsules is one of the key factors for keeping their physical integrity and stability in the course of their synthesis process, storage and handling operations to fabricate a coating. In this work, the melamine-formaldehyde resin was chosen to form the shell material due to its good thermo-mechanical and water-resistant properties. A comparison between the syntheses of microcapsules based on melamine-formaldehyde and urea-formaldehyde resins is included. Selection of emulsifiers is another key factor because the microcapsules formation is influenced by the emulsion stability. Therefore, a series of experiments was conducted changing the emulsifying system and a discussion of results is provided. The synthesis products were examined by optical microscopy and scanning electronic microscopy (SEM). Further characterization applied to a powder of microcapsules allowed to verify the effective repairing agent encapsulation.

Chapter 4 - In this approach, a sensitive chemical sensor was developed to detect melamine selectively by electrochemical approach, where ternary mixed metal oxide ($ZnO/CuO/Co_3O_4$) nanoparticles (NPs) were prepared by wet-chemical process. The calcined $ZnO/CuO/Co_3O_4$ NPs were investigated by field emission scanning electron microscopy (FESEM), energy-dispersive X-ray spectroscopy (EDS), X-ray photoelectron spectroscopy (XPS), powder X-ray diffraction (XRD), ultraviolet visible spectroscopy (UV-vis), and Fourier-transform infrared spectroscopy (FTIR). To the fabricate melamine sensor, slurry of NPs in ethanol was deposited as uniform thin layer on a glassy carbon electrode (GCE). The calibration curve of the proposed melamine sensor in form of current versus concentration (in logarithmic scale) plot is found to be linear over a melamine concentration range of 0.05 nM ~ 0.05 mM. The sensitivity of the sensor is very good and detection limit is very low. The melamine sensor with active ZnO/CuO/Co$_3$O$_4$ NPs shows good reliability, precise reproducibility and short response time in sensing performances. The developed ternary metal oxide nanoparticles based sensor is a new introduction in the sensor technology for determining melamine in environmental samples reliably.

Chapter 5 - Melamine and cyanuric acid had been implicated in kidney-related diseases in infants and in the death of a large number of cats and dogs that ingested tainted food containing melamine. These incidents were caused by the willful adulteration of the raw ingredients with melamine in the dairy products and pet food, respectively, in order to boost the apparent protein content in the nutritional labels. Melamine and cyanuric acid can form extremely insoluble particles, which are composed of hydrogen-bonded melamine-cyanuric acid complex or melamine-cyanurate. Ingested melamine and cyanuric acid are both absorbed in the gastrointestinal tract, distribute systemically, and precipitate as the melamine-cyanurate complex in the renal tubules, leading to progressive tubular blockage, degeneration, and acute renal failure. Melamine ingestion has also been implicated in neurological and reproductive toxicity. Due to the very active research and development of screening techniques aimed at reducing the incidence of melamine food contamination, the risk of renal failures due to melamine has been curtailed significantly. However, there are other routes of exposures to melamine and cyanuric acid that would still contribute to the adverse human health effects. The U. S. Food and Drug Administration (FDA) reported that melamine is incorporated into melamine-formaldehyde resins for making food packaging materials, plastic tableware, and the coating of cans in canned foods. Consequently, food and beverage products have been found to contain melamine at trace levels as a result of leaching from melamine-containing resins. Trace levels of cyanuric acid can be present in food and water from the use of dichloroisocyanurate in drinking water, swimming pools, and water used in food manufacturing. Cyanuric acid derivatives are also found in sanitizing solutions for food processing equipment and utensils.

This chapter will evaluate the formation of mealime-derived particles in aqueous and biological matrices using different analytical techniques for studying the bioaccumulation of melamine-cyanurate in tissues including kidney stones. The histomorphologic characteristics of the crystals formed at various concentrations and temperatures can be charatertized using scanning electron microscopy (SEM) to determine the crystallite morphology, size, and distribution. The authors' data indicated that the melamine-cyanurate crystals produced at 37°C were coarser and larger compared to those formed at 25°C at 100 ppm levels. Furthermore, the proportion of "spoke-like" crystals decreased along with the accompanying increase in the proportion of "needle-like" crystals at the higher temperature of 37°C. Both bovine serum albumin and polyvinylpyrrolidone, a synthetic macromolecule, have been found to alter the crystal morphology to a spherical form, which is typically observed for the particles in the kidney microtubules. Samples containing melamine-cyanurate formed in bovine blood plasma and in the kidney tissue of catfish that had been fed daily for 3 days with 200 milligram per day of melamine-cyanuric acid complex per kilogram of body weight were also analyzed by SEM and Raman microscopy. Other researchers have used X-ray diffraction (XRD), SEM with energy dispersive X-ray microanalysis, and Fourier transform IR spectroscopy to classify urinary stones. The results show that oxalates (43.3%) are the most common followed by phosphates (13.3%), urates (6.7%), and mixed stone (36.7%).

Chapter 6 - In this approach, nanostructure materials (Cadmium doped antimony oxide; CAO NSs) were synthesized by a facile wet-chemical method at a low temperature to detect melamine from aqueous solution. The calcined CAO-NSs were characterized systematically by FE-SEM, EDS, UV/Vis., FTIR spectroscopy, powder XRD, and XPS techniques. The glassy carbon electrode (GCE) was modified with the CAO-NSs and sensing performance towards the selective melamine was explored by the electrochemical approach in phosphate buffer solution. The melamine undergoes a reduction reaction in the presence of CAO-NSs/GCE in PBS.

Preface

The CAO-NSs/GC electrode attained higher sensitivity for a wide range of concentration, ultra-low limit of detection, long-term stability, excellent repeatability, and reproducibility. This method might represent an efficient way of sensitive sensor development for the determination of toxic melamine and their derivatives in broad scales.

In: An Introduction to Melamine
Editor: Ashley Harris

ISBN: 978-1-53617-136-5
© 2020 Nova Science Publishers, Inc.

Chapter 1

MELAMINE AND ITS ANALOGOUS: ANALYTICAL METHODS AND RELATED LEVELS IN FOODSTUFFS

Mena Ritota[*] *and Pamela Manzi*
CREA – Centro di ricerca Alimenti e Nutrizione, Rome, Italy

ABSTRACT

In the past decade, adulteration incidents concerning the fraudulent addition of melamine in dairy products and vegetable proteins occurred in China prompted numerous governmental and private laboratories to develop analytical methods for melamine and related compounds determination in foods. Since then, many efforts have been made to develop both screening and confirmatory methods based on traditional analytical techniques (liquid and gas chromatography, mass spectrometry, liquid and gas chromatographies coupled to mass spectrometry, capillary electrophoresis) and increasingly innovative techniques (i.e., immunoassay, sensor technology, vibrational spectroscopy).

In this chapter, technical and critical aspects concerning confirmatory methods for the analysis of melamine and its analogous, including cyanuric

[*] Corresponding Author's Email: mena.ritota@crea.gov.it.

acid, ammeline, and ammelide, in foodstuffs will be discussed. Moreover, an overview of the concentrations of melamine and related compounds reported in the recent literature for different food items (i.e., dairy products, meat, seafood, egg, vegetables, soy and cereal products) will be summarized, as the results from monitoring these compounds in food products indicate their continuing presence in the food chain. Sources of baseline melamine and cyanuric acid in foodstuffs, in fact, may come from melamine-containing plastic materials for tableware, herbicide/pesticide and fertilizers use, or from the industrial production of melamine resins, which may be an ongoing source of melamine in water and the environment in general. At the same time, melamine can be degraded *via* deamination reactions to its analogues ammeline, ammelide, and cyanuric acid.

Melamine and cyanuric acid are the major compounds observed in foods, ammeline and ammelide generally accounting for less than 10% of the total content. Fortunately, current melamine levels reported in foods are well below the tolerance limits set by Food and Drug Administration (2.5 mg/kg and 1.0 mg/kg in milk products and powdered infant formula, respectively).

Keywords: melamine, analytical methods, foodstuffs

1. INTRODUCTION

Melamine (2,4,6-triamino-1,3,5-triazine, CAS number 108-78-1) (MEL) is a heterocyclic compound with high nitrogen content (about 66% by mass). Thanks to this characteristic and its low price, in the past it was fraudulently added to pet feed and foods in order to increase their apparent protein content. Other nitrogen rich compounds often involved in milk adulteration are whey and urea. Whey is the by-product of cheese manufacturing that is added to liquid milk not only to increase the protein content but also its volume, and is a very cheap product, while urea, naturally present in milk in low concentration, has been extensively used in frauds to increase the apparent content of milk, due to its low cost (Nascimento et al. 2017). Melamine contamination, however, resulted in several illness and death in pets, as well as in infants and young children, and since than the adulterant MEL has become an important issue attracting worldwide attention.

Nowadays, the sources of melamine contamination in food are generally devided into adulteration levels and baseline levels. The first ones refer to levels of melamine derived from the intentional and illegal addition of melamine, while baseline levels are defined as levels of melamine that do not result from adulteration or misuse (Hilts and Pelletier 2009). These are expected to be <1 mg/kg, and they are not considered to be a health concern (Hilts and Pelletier 2009). Therefore, low levels of melamine are generally found in foods, not due to contamination but to normal food production and processing, such as migration from food contact materials, pesticides or fertilizer use (Gossner et al. 2009). These levels are generally in the microgram per kilogram range (Gossner et al. 2009).

Infant formulae and dairy products are the most common foods analyzed for melamine content, but MEL is also reported in marine products, since it may be added to the binding agents of some fish and crustacean pellet feeds (Tittlemier, Lau, Menard, Corrigan, Sparling, Gaertner, Cao, Dabeka, et al. 2010). Also food items containing vegetables on which cyromazine (CYRO) may be applied as an insecticide, such as mushrooms, spinach, potatoes, onion, and tomatoes, have to be analyzed for their melamine content (Tittlemier, Lau, Menard, Corrigan, Sparling, Gaertner, Cao, Dabeka, et al. 2010), since MEL is a degradation product of the pesticide cyromazine (Arnold 1990). In some cases, MEL and its analogous are also investigated in bakery products (bread, biscuits, cakes), because they are largely consumed food items and may contain MEL residues derived from the potential incorporation of wheat gluten, which showed to be MEL contaminated in 2007 (Zhu and Kannan 2018). Finally, also food products of animal origin, such as meat and eggs, are analyzed for their melamine levels due to the potential transfer of MEL from contaminated animal feedstuff to food items. It has been reported, in fact, that laying hens fed with MEL contaminated feeds showed MEL residues both in eggs (Valat et al. 2011, Wang et al. 2012) and tissues (Valat et al. 2011), with a high dose-dependent effect particularly noticeable in eggs (Valat et al. 2011).

Microbial metabolism of melamine may result in the production of its analogous (Figure 1), cyanuric acid (CYA), ammeline (AMN), and ammelide (AMD), which are also the by-products of melamine manufacturing (Miao et al. 2009). Furthermore, degradation of s-triazine pesticides, such as simazine and atrazine, products cyanuric acid (Karbiwnyk et al. 2009), which can be also used as microbicide and disinfectant in water treatment applications as a chlorine stabilizer (Kowalsky 1992). Cyanuric acid is also an FDA-accepted component of feed-grade biuret, a ruminant feed additive (Food and Drug Administration 2008).

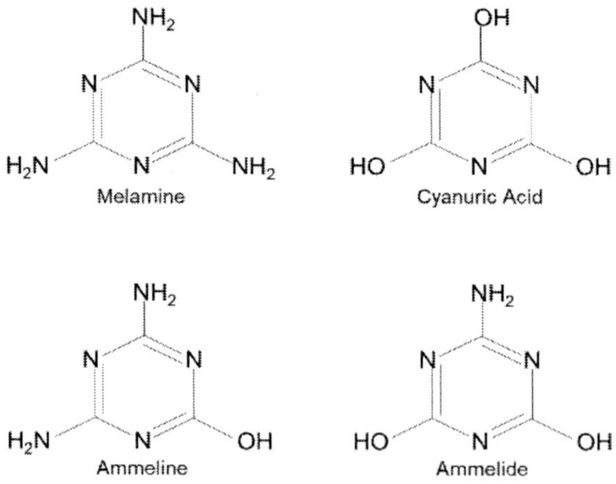

Figure 1. Structure of melamine (MEL), cyanuric acid (CYA), ammeline (AMN), and ammelide (AMD).

Aside from melamine contents in foodstuffs due to adulteration, world population is continuously exposed to baseline concentrations of MEL, due to its widespread presence in the environment and food chain. Therefore, this work focuses on the levels of melamine and its analogous (CYA, AMN and AMD) in foods, in order to provide information on the extent of human exposition to MEL *via* the food consumption. Migration of melamine from packaging materials to foods and melamine contaminated feeds are not discussed in this chapter.

Furthermore, an overview of the confirmatory methods available for MEL and its derivatives detection is provided, with attention focused on the theoretical and practical aspects of these techniques.

2. ANALYTICAL METHODS

In 2008 the fraudulent addition of melamine to foods for artificially increasing their protein content, which had been cause of illness and death of several infants and young children, brought into question the problem of determining the actual protein levels in foods deriving from protein-based nitrogen, in order to ensure food quality and safety.

The Kjeldahl method, based on nitrogen content measures, is the official analytical method to determine total protein in foodstuffs (AOAC 2005). Indeed, the Kjeldahl method does not discriminate between endogenous nitrogen from protein sources and exogenous nitrogen fraudulently added, therefore it produces unacceptable results for adulterated samples with non-protein nitrogen (Moore et al. 2010). In the same way, other analytical methods determining protein content in foods, such as the Dumas and the modified Lassaigne methods, are not able to distinguish protein-based nitrogen from non-protein nitrogen (Domingo et al. 2014).

After the incidence of melamine contamination, several analytical methods have been developed to determine MEL, as well as its related compounds, in foodstuffs, primarily in milk and dairy products, where MEL contamination first occurred, and then in many others food products. Since several scientific works have already reported the use of different analytical techniques for detecting and quantifying MEL and its analogous in several food matrices (Domingo et al. 2014, Nascimento et al. 2017, Ritota and Manzi 2017, Rovina and Siddiquee 2015, Wang et al. 2016), in this chapter only standard methods for detecting MEL and its analogous in foods will be discussed.

2.1. Sample Preparation

When using selective analytical techniques, sample preparation procedures are very simple and generally consist of liquid extraction followed by a purification step, if needed (Tittlemier 2010).

Melamine and its analogous are small polar molecules, therefore liquid extraction is generally performed with polar solvents (Tittlemier 2010).

Standard method (ISO-IDF 2010) extracts melamine and its related compound with an organic solvent, or a mixture with water, while simultaneously achieving protein precipitation. MEL, in fact, could bond with proteins through some not specified interactions (Sun et al. 2010), therefore it is recommended to remove proteins before melamine analysis, especially in food items containing very high levels of proteins, such as milk and dairy products. In more detail, LC-MS/MS method for melamine and cyanuric acid determination in milk, milk products and infant formulae (ISO-IDF 2010) extracts MEL and CYA with an acetonitrile:water solution (35:15, v/v) or with acetonitrile alone, and sonicate and/or centrifuge the obtained solution before injecting the supernatant into the HPLC system. The same method also specifies that *"Extraction may be done with a minimum of 5 milliliters of extraction solvent per gram of sample material, whereby the final percentage organic solvent in the extraction mixture should be minimal 70%"* and that *"Extraction should take place for at least 5 min by either vigorous shaking or a combination of ultrasonification and vortexing"* (ISO-IDF 2010). In the determination of melamine, ammeline, ammelide and cyanuric acid by GC-MS method (Litzau, Mercer, and Mulligan 2008), instead, a mixture composed of diethylamine:water: acetonitrile (10:40:50, v/v) is used for the extraction procedure, followed by sonication, centrifugation and filtration (0.45 μm) before sample derivatization. The introduction of diethylamine into the solvent system not only allows the extraction of all four compounds (ammeline and ammelide have low solubility in traditional extraction solvents), but also allows the disruption of the hydrogen bonds of the supramolecular aggregates which form with melamine in the presence of cyanuric acid, and that could be cause of recovery loss due to the insolubility of the melamine-cyanurate complexes (Litzau, Mercer, and Mulligan 2008).

Also a mixture of organic solvents with acidic solutions can be used for the simultaneous extraction of analytes and protein precipitation from the matrix (Ritota and Manzi 2017). Andersen et al. (2008), for example, observed an increase in the extraction recoveries of MEL from fish tissue resulted from acidifying the sample in the initial aqueous acetonitrile extraction by hydrochloric acid (HCl). Also Tittlemier et al. (2009) observed an improvement in the extraction efficiency of MEL from infant formula by using diluted HCl, since it helped to separate the phases during centrifugation. Sun et al. (2010), instead, employed a trichloroacetic acid (TCA) solution to precipitate proteins and to dissociate MEL from the sample matrix (liquid milk), and improved extraction efficiency by adding 2% lead acetate to 1% trichloroacetic acid (1:25, v/v). Also Meng et al. (2015) employed TCA (20%) to precipitate proteins from infant formula samples and to dissociate analytes from the matrix, but in a solution with acetonitrile (1:100, v/v), in order to simultaneously extract cyromazine, melamine, ammelide, ammeline, cyanuric acid, and dicyandiamide. The same authors (Meng et al. 2015) also observed a decrease in the analytes recovery when TCA ratio exceeded 2%, probably due to the pH of the mixed solution, which was less than 4.

When using analytical methods with low limits of detection, more complex purification procedures are needed for MEL and its analogous determination, in order to avoid matrix interferences (Tittlemier 2010). In addition to protein removal, sample defatting by relatively non polar solvents (n-hexane or dichloromethane) could be helpful to remove co-extracted lipids (Tittlemier 2010). A further clean-up step, after initial extraction, is generally performed by solid phase extraction (SPE). Being melamine a weakly alkaline compound, cation-exchange/reversed phases are the most used sorbents to purify samples and enrich the melamine extracts (Ritota and Manzi 2017), while cyanuric acid is generally purified and enriched on anion-exchange/reversed phases SPE cartridges (Smoker and Krynitsky 2008, Braekevelt et al. 2011). Ammeline and ammelide are amphoteric compounds and can be retained by cation-exchange as well as by anion-exchange/reversed phases SPE cartridges, as observed by (Braekevelt et al. 2011). The same authors (Braekevelt et al. 2011) also reported a very low recovery of ammelide and ammeline from the Oasis MCX SPE cartridges, unless the applied aliquot contained 0.05 g or less of sample (infant formula). AMN and AMD were better eluted from the same cartridges by using 5% NH_4OH in MeOH, unlike the basic acetonitrile solution used by Smoker and Krynitsky (2008) for eluting melamine from the Oasis MCX SPE cartridges. The cation-exchange (for MEL, AMN and AMD clean-up) and the anion-exchange (for CYA clean-up) extracts have to be analyzed separately in order to minimize interferences, even if Braekevelt et al. (2011) obtained sufficient chromatographic separation between CYA and the other analytes so to determine all the compounds in the same MS run; in such a way, the two fractions can be combined before evaporation and run as a single sample. In order to improve selectivity and reduce time analysis of SPE, new sorbents such as molecularly imprinted polymers (MIP) have been developed for the analysis of MEL and its related compounds (Ge et al. 2015).

Attention must be paid to the homogenization of the sample. For example, when analyzing vegetable products, Tittlemier, Lau, Menard, Corrigan, Sparling, Gaertner, Cao, Dabeka, et al. (2010) found larger variations in MEL recoveries in spiked canned tomatoes compared to $^{13}C_3$-MEL internal standard recoveries, ranging between 34 and 102% and 83-86%, respectively. The authors (Tittlemier, Lau, Menard, Corrigan, Sparling, Gaertner, Cao, Dabeka, et al. 2010) ascribed this variation to the incomplete initial homogenization of the samples, since different levels of MEL were found in the various parts of the sample, with significantly higher concentrations in the seeds than in the liquid and pulp of canned tomatoes (one-way ANOVA; $p<0.001$). According to the study of Root, Hongtrakult, and Dauterman (1996), in fact, the pesticide cyromazine is absorbed and translocated in tomato plant, with the consequent production of melamine. The problem solving was the initial homogenization of the canned tomato samples in a food processor and a further use of a hand-held homogenizer during the first extraction step: in this way the average MEL recovery coefficient of variation reduced from 44% to 5% (Tittlemier, Lau, Menard, Corrigan, Sparling, Gaertner, Cao, Dabeka, et al. 2010).

2.2. Standard Analytical Techniques

When choosing an analytical method for analysis of food contaminant, regulatory limits must be considered in order to reach the optimum method sensitivity. This is defined as *"the lowest analyte concentration in the matrix that can be measured with acceptable accuracy and precision (i.e., LLOQ)"* (Food and Drug Administration 2018) or, according to the EURACHEM Guide 2014 (Eurachem 2014), as *"the change in instrument response which corresponds to a change in the measured quantity (for example an analyte concentration), i.e., the gradient of the response curve"* (BIPM 2012).

Regarding melamine, its maximum residual levels have been set at 1 mg/kg (Joint FAO/WHO Expert Committee on Food Additives 2010) and 0.15 mg/kg in powdered and liquid infant formula, respectively (FAO 2012), while the content allowed in other foods and animal feed is 2.5 mg/kg (FAO 2010).

The required limit of quantification of an analytical method should be at least 10-fold lower than the maximum limit (ML) of MEL to ensure reliable quantification at ML (ISO-IDF 2010). Several selective quantitative methods reported in the literature seem to be sensitive enough to detect MEL at 1.0 and 2.5 mg/kg (Tittlemier 2010, Ritota and Manzi 2017), the maximum residual level of MEL for powdered infant formula and other foods (Joint FAO/WHO Expert Committee on Food Additives 2010), respectively, but the limits of quantification reported in the literature are often calculated in different ways, so they are not comparable (Tittlemier 2010). Furthermore, some of the information regarding methods validation are not always available (Tittlemier 2010), so the applicability of these methods could be questioned.

Liquid chromatography tandem mass spectrometry has proven to be a high specificity and sensitivity method for MEL detection (Ritota and Manzi

2017) and it is used as the international standard method for melamine and cyanuric acid determination in milk, milk products and infant formulae (ISO-IDF 2010). The method allows to reach a limit of quantification (LOQ) of 0.05 mg/kg for MEL and 0.10 mg/kg for CYA, both in cow milk and milk-based powdered infant formula.

In general, high performance or ultra performance liquid chromategraphy (HPLC or UPLC) and gas chromatography (GC) tandem mass spectrometry (MS/MS) give good selectivity thanks to the capability of chromatography to separate analytes from co-extracted materials and the use of multiple reaction monitoring during detection (Tittlemier 2010). Other detectors, such as single-stage mass spectrometry (MS), diode array detection (DAD), and ultraviolet absorption (UV) give lower selectivity than MS/MS detector (Hilts and Pelletier 2009).

The common use of liquid chromatography for analysis of MEL and its analogous is due to the small and polar nature of the molecules (Ritota and Manzi 2017). However, due to their characteristics, MEL and related compounds showed poor retention and/or separation on the reversed phase (RP) columns (Ehling, Tefera, and Ho 2007, Gratz, Gamble, and Heitkemper 2009, Sancho et al. 2005). Some authors used ion-pair reagents to improve separation on RP columns (Sancho et al. 2005, Sun et al. 2010), but this could result in UV adsorption decrease (Ishiwata et al. 1987) or ionization suppression in the MS source (Sancho et al. 2005). Further attempts to achieve sufficient retention and improve separation in RP-LC employed porous graphitic carbon columns for analysis of CYA in fish and shrimp sample extracts (Karbiwnyk et al. 2009, Pichon et al. 1995). The last improvements in the field of chromatographic phases for LC columns resulted in the development of hydrophilic interaction liquid chromatography (HILIC), which is still used in the international standard method for MEL and CYA determination in milk products, alone or with zwitterionic functional groups (ZIC-HILIC) (ISO-IDF 2010).

Due to the polar nature of the compounds, capillary electrophoresis could be a good alternative to liquid chromatography for the separation of MEL and its related compounds (Chen and Yan 2009). CE provides high separation efficiency, high speed, and low consumption of solvent and sample (Ritota and Manzi 2017). However, it suffers for the lack of sensitivity and low reproducibility (Xia et al. 2010). Furthermore, the difficulty of obtaining an appropriate standard reference for MEL analysis by CE, which could result in unambiguous identification, implies the need to use hyphenated methods, such as CE-MS or LC-MS (Vallejo-Cordoba and Gonzalez-Cordova 2010).

While MEL and its analogous can be analyzed by LC as such, when using GC for their determination a derivatization step, generally through the formation of trimethylsilyl (TMS) derivatives, is required, since MEL and its related analytes are relatively involatile (Tittlemier 2010). However, MEL with three NH_2 groups in the structure can form different derivative products (mono-, di- and three-TMS-melamine derivatives), which can result in poor reproducibility of the GC method (Xu et al. 2009). Derivatization step also implies long analysis time and cost of the reagents (Ritota and Manzi 2017). Furthermore, the presence of water in the sample to be derivatized prevents the formation of TMS derivatives of the analytes (Litzau, Mercer, and Mulligan 2008): problems associated with the derivatization step can be identified by the lower response than usual (<30%) of the internal standard, or if the vial significantly warms to the touch after the addition of the derivatization reagent (Litzau, Mercer, and Mulligan 2008). All these disadvantages limit the use of GC as separation technique for MEL analysis, even if a GC-MS method was developed by FDA for the determination of melamine, ammeline, ammelide and cyanuric acid in different matrices (Litzau, Mercer, and Mulligan 2008). The method establishes a minimum reporting level of 10 µg/g of each target analyte, and also reports a guidance for screening all four analytes in foods at a minimum reporting level of 2.5 µg/g (switching the detection from full scan mode to single ion monitoring), but with the method performances that must be verified in each laboratory (Litzau, Mercer, and Mulligan 2008). The same method also highlights the difficulties in derivatizing the analytes (particularly cyanuric acid, and in a less extent ammeline and ammelide) in the presence of high sugar sample matrix: in that case, the addition of further derivatization agent may improve the analytes recovery.

Miao et al. (2009) modified and optimized the method developed by FDA (Litzau, Mercer, and Mulligan 2008) in order to simultaneously detect MEL, AMN, AMD, and CYA in milk and milk products by GC-MS/MS. The use of a mixture of pyridine and acetonitrile (1:1, v/v) for re-dissolving the extracts before derivatization enhanced the response of the four compounds by 3-folds compared to acetonitrile or pyridine alone, and the use of tandem mass spectrometry allowed to increase sensitivity (LOQ = 0.005 mg/kg for all compounds, with a ratio of signal to noise of 10). Furthermore, the authors (Miao et al. 2009) used a more polar column (VF-5ms quartz capillary column (30 m×0.25 mm i.d. × 0.25 μm)) than that employed by the FDA method (DB-5MS (30 m × 0.25 mm i.d. × 0.25 μm), and increased the final temperature of the GC oven (from 270°C to 300°C), as well as the holding time (from 2 to 10 minutes), in order to eliminate the interference coming from the last injection and improve separation.

Some authors (Zhang, Ma, and Fan 2014) also employed a microwave-assisted derivatization to shorten the time consuming pre-treatment step in the analysis of melamine in fish, shrimp, clam, and winkle by GC-MS.

To overcome the problem of derivatization, Xu et al. (2009) developed a method for the direct detection of melamine in dairy products by GC/MS. Due to its polar nature, MEL should be detected by strong GC polar columns (such as polyethylene glycol column), but they suffer from the problem of the column bleeding (Xu et al. 2009). Therefore, the authors coupled a weak polar column DB-5MS (30 m (5%-phenyl)-methylpolysiloxane, 0.25mm i.d., 0.25 μm df) with a short length of strong polar column Innovax (polyethylene glycol, 0.32mm i.d., 0.25 μm df), in order to avoid the band broadening and peak tailing of the less polar column (DB-5MS).

When mass spectrometry is employed as detection technique, MEL is generally analyzed in the positive electrospray ionization mode (ESI$^+$) and CYA in the negative electrospray ionization mode (ESI$^-$) (Tittlemier 2010). A well defined chromatographic separation is needed if MEL and CYA are simultaneously analyzed, by switching polarity during the same run, otherwise they have to be detected separately (Smoker and Krynitsky 2008).

According to the Commission Decision 2002/657/EC (Commission of the EU Communities 2002), monitoring at least two transitions is recommended for MS^n analysis. The most appropriate transition (most intense) shall be used for quantification, while the second one for confirmation (ISO-IDF 2010). In the case of melamine, the most often transitions monitored are 127 → 85 and 127 → 68, which correspond to a loss of cyanamide from the [M+H]$^+$ ion and a subsequent loss of ammonia (Varelis and Jeskelis 2008). Generally, the former transition reaction is used for MEL quantification, while the latter for analyte confirmation (ISO/TS 15495). Regarding CYA, the transitions monitored are 128 → 42 and 128 → 85 (ISO-IDF 2010), which correspond to the loss of the neutral fragment 2,4-diamino-1,3-diazete from the molecular ion, to produce the conjugate acid of cyanamide (m/z 43), and to the loss of cyanamide from the [M-H]$^-$ ion (Varelis and Jeskelis 2008). Generally, the former transition reaction is used for CYA quantification, while the latter for analyte confirmation. The standard method (ISO-IDF 2010) also defines the other confirmation criteria for mass spectrometric detection of the target analytes, according to the what established by the Commission Decision 2002/657/EC (Commission of the EU Communities 2002): i) signal visible at the two diagnostic transition reactions selected for each analyte and at the two diagnostic transition reactions selected for its corresponding IS, with a signal to noise ratio for each diagnostic ion ≥ 3:1; ii) relative retention time of the analyte (ratio of the chromatographic retention time of the analyte to that of the corresponding IS) corresponding to that of the averaged retention time of the calibration solutions within a well defined tolerance; iii) peak area ratio from the different transition reactions recorded for each analyte is within a well defined tolerance.

When mass spectrometry is employed for MEL analysis, matrix effects are observed as suppression of analyte mass spectrometric response, with suppression decreasing with MEL concentration of sample matrix decrease (Tittlemier, Lau, Menard, Corrigan, Sparling, Gaertner, Cao, Dabeka, et al. 2010). Furthermore, a variation in the magnitude of matrix effects was observed not only among different food items matrices (Tittlemier, Lau, Menard, Corrigan, Sparling, Gaertner, Cao, Dabeka, et al. 2010), but also among similar food products matrices (Tittlemier et al. 2009). This difference implies the need to use stable isotope-labelled internal standards to correct for matrix effects in melamine analysis by mass spectrometry (Tittlemier, Lau, Menard, Corrigan, Sparling, Gaertner, Cao, Dabeka, et al. 2010). Samples should be fortified with stable isotope-labelled internal standards before extraction, in order to take into account the loss of analyte during sample preparation (Tittlemier, Lau, Menard, Corrigan, Sparling, Gaertner, Cao, Dabeka, et al. 2010). In the case in which dilution makes pre-extraction fortification not feasible, sample extracts should be fortified before MS analysis (Tittlemier, Lau, Menard, Corrigan, Sparling, Gaertner, Cao, Dabeka, et al. 2010).

However, Xia et al. (2009), when analyzing egg samples fortified with high level of melamine, e.g., 1 mg/kg, observed an interfering peak at the retention time of melamine-$^{15}N_3$, which resulted from the isotopic ion of melamine. Since the recoveries obtained by the standard calibration curve were acceptable (>80%), the authors performed MEL GC-MS analysis by external calibration without using melamine-$^{15}N_3$.

3. LEVELS OF MELAMINE AND ITS ANALOGOUS IN FOODSTUFFS

3.1. Melamine Crisis: From 2008 until 2011

Following the milk and milk products adulteration with melamine occurred in China in 2008 (Gossner et al. 2009), a lot of worldwide investigations were carried out for monitoring melamine, and in some cases its analogous, in milk, milk ingredients and composited products containing milk-derived ingredients. Data reported all over the world indicated high levels of melamine in a lot of food items (Hilts and Pelletier 2009), with melamine levels that in some cases exceeded 6500 mg/kg (Table 1).

In 2009 the Institute of Nutrition and Food Safety, Chinese Center for Disease Control and Prevention (Wu, Zhao, and Li 2009) reported data concerning 111 infant formula collected in 2008 in Beijing and Gansu provinces from Sanlu Company (one of the largest milk powder manufacturers in China and implicated in the contamination event): 87 products resulted positive for melamine adulteration and ranged between 118 and 4700 mg/kg. Very high levels of MEL contamination were also observed in the raw material analyzed from Gansu province and used for milk adulteration (Wu, Zhao, and Li 2009): MEL, in fact, had been added to the raw ingredients at the milk collection centers, in order to increase milk apparent protein content after dilution with water. In the same work, the authors (Wu, Zhao, and Li 2009) reported data concerning 38 infant formula collected from other companies (non-Sanlu) in Gansu province: all products had very lower melamine concentration compared to Sanlu infant formula (4.06 *vs* 1673.6 mg/kg, reported as mean value). These results agreed with those reported by. China's General Administration of Quality Supervision, Inspection and Quarantine (AQSIQ) in mid-September 2008, according to which MEL levels in different Chinese infant formula (n = 22) were in the range (0.09–2563) mg/kg (Hilts and Pelletier 2009). Melamine adulteration, instead, was not observed in infant formula sold in Canada between September 22 and October 6, 2008: Tittlemier et al. (2009), in fact, analyzed 94 products purchased from the major retailers in Ottawa and reported MEL values in the range (0.00431 - 0.346) mg/kg. The same infant formulae were analyzed later by Braekevelt et al. (2011) for their ammeline, ammelide and cyanuric acid content: CYA was detected in almost all infant formula samples (96.7%) and had the highest concentration (maximum level observed 0.45 mg/kg), while AMN and AMD contents were very low (< 0.02 mg/kg). In general, powdered infant formula showed higher level of MEL and CYA compared to liquid formula. However, analytes ratios (MEL/CYA, MEL/AMN e MEL/AMD) significantly differed from those reported by Wu, Zhao, and Li (2009), indicating a different source of melamine and analogous contamination (Braekevelt et al. 2011). Cyanuric acid was present in higher concentration than MEL also in the human-grade vegetable protein products analyzed by Levinson and Gilbride (2011).

Table 1. Melamine crisis: from 2008 until 2011

Matrix	Melamine range	N. Samples	Positive Samples	References
U.S. Market-ready fish	< 0.0047 - 0.237 mg/kg	111	33	(Andersen et al. 2008)
Worldwide biscuits, cakes and confectionary collected in 2008	0.6 - 945.86 mg/kg	NR	115	(Hilts and Pelletier 2009)
Worldwide liquid milk and yoghurt collected in 2008	< 2.5 - 648 mg/kg	NR	12	(Hilts and Pelletier 2009)
Worldwide snack foods collected in 2008	0.5 - 54 mg/kg	NR	14	(Hilts and Pelletier 2009)
Worldwide powdered and cereal products collected in 2008	0.38 - 1143 mg/kg	NR	5	(Hilts and Pelletier 2009)
Worldwide processed foodstuff collected in 2008	0.6 - 14 mg/kg	NR	8	(Hilts and Pelletier 2009)
Worldwide food-processing ingredients collected in 2008	2.8 - 6694 mg/kg	NR	4	(Hilts and Pelletier 2009)
Chinese infant formulae (collected from Sanlu Company in 2008)	< 0.05 - 4700 mg/kg	111	87	(Wu, Zhao, and Li 2009)
Chinese infant formulae (collected from other companies in 2008)	4.06 mg/kg	38	NR	(Wu, Zhao, and Li 2009)[1]
Biscuits, cakes and confectionery collected in 2008 in China	3.2 - 68 mg/kg	NR	22	(Hilts and Pelletier 2009)
Liquid milk and yoghurt collected in 2008 in China	0.765 - 9.9 mg/kg	NR	30	(Hilts and Pelletier 2009)
Snack food collected in 2008 in China	18 mg/kg	NR	1	(Hilts and Pelletier 2009)
Frozen dessert collected in 2008 in China	4.4 - 21 mg/kg	NR	5	(Hilts and Pelletier 2009)
Infant formula collected in 2008 in China	0.09 - 2563 mg/kg	NR	22	(Hilts and Pelletier 2009)
Powdered milk and cereal products collected in 2008 in China	0.53 - 6196.61 mg/kg	NR	37	(Hilts and Pelletier 2009)
Food-processing ingredients (egg powder) collected in 2008 in China	0.1 - 4 mg/kg	NR	4	(Hilts and Pelletier 2009)
Other products (fresh eggs) collected in 2008 in China	2.9 - 4.7 mg/kg	NR	3	(Hilts and Pelletier 2009)
Infant formulae (sold in Canada from 22/09 to 6/10/2008)	< 0.004 - 0.346 mg/kg	94	71	(Titlemier et al. 2009)[2]
Candy (sold in Canada between 30/09 and 2/10/2008)	< 0.004 - 7.29 mg/kg	11	3	(Tittlemier, Lau, Menard, Corrigan, Sparling, Gaertner, Cao, and Dabeka 2010)[2]
Cheese, soft (sold in Canada between 30/09 and 2/10/2008)	< 0.004 mg/kg	2	0	(Tittlemier, Lau, Menard, Corrigan, Sparling, Gaertner, Cao, and Dabeka 2010)[2]

Matrix	Melamine range	N. Samples	Positive Samples	References
Cookies (sold in Canadian retail outlets between 30/09 and 2/10/2008)	< 0.004 mg/kg	1	0	(Tittlemier, Lau, Menard, Corrigan, Sparling, Gaertner, Cao, and Dabeka 2010) [2]
Custard (sold in Canada between 30/09 and 2/10/2008)	0.0160 mg/kg	1	1	(Tittlemier, Lau, Menard, Corrigan, Sparling, Gaertner, Cao, and Dabeka 2010) [2]
Ice cream (sold in Canada between 30/09 and 2/10/2008)	< 0.004 mg/kg	1	0	(Tittlemier, Lau, Menard, Corrigan, Sparling, Gaertner, Cao, and Dabeka 2010) [2]
Milk (sold in Canada between 30/09 and 2/10/2008)	0.00742 mg/kg	47	1	(Tittlemier, Lau, Menard, Corrigan, Sparling, Gaertner, Cao, and Dabeka 2010) [2]
Milk, condensed and evaporated (sold in Canada between 30/09 and 2/10/2008)	< 0.004 - 0.0307 mg/kg	12	4	(Tittlemier, Lau, Menard, Corrigan, Sparling, Gaertner, Cao, and Dabeka 2010) [2]
Milk, powder (sold in Canada between 30/09 and 2/10/2008)	0.00528 - 0.0122 mg/kg	2	2	(Tittlemier, Lau, Menard, Corrigan, Sparling, Gaertner, Cao, and Dabeka 2010) [2]
Milk-containing beverage (sold in Canada between 30/09 and 2/10/2008)	< 0.004 - 0.282 mg/kg	35	11	(Tittlemier, Lau, Menard, Corrigan, Sparling, Gaertner, Cao, and Dabeka 2010) [2]
Milk-containing cereal mix (sold in Canada between 30/09 and 2/10/2008)	< 0.04 mg/kg	1	0	(Tittlemier, Lau, Menard, Corrigan, Sparling, Gaertner, Cao, and Dabeka 2010) [2]
Milk-containing drink mix (sold in Canada between 30/09 and 2/10/2008)	< 0.004 - 0.0213 mg/kg	12	3	(Tittlemier, Lau, Menard, Corrigan, Sparling, Gaertner, Cao, and Dabeka 2010) [2]
Soy-based cheese substitute (sold in Canada between 30/09 and 2/10/2008)	< 0.004 mg/kg	4	0	(Tittlemier, Lau, Menard, Corrigan, Sparling, Gaertner, Cao, and Dabeka 2010) [2]
Soy-based sour cream substitute (sold in Canada between 30/09 and 2/10/2008)	< 0.004 mg/kg	1	0	(Tittlemier, Lau, Menard, Corrigan, Sparling, Gaertner, Cao, and Dabeka 2010) [2]
Soy-containing beverage (sold in Canada between 30/09 and 2/10/2008)	0.00663 mg/kg	38	1	(Tittlemier, Lau, Menard, Corrigan, Sparling, Gaertner, Cao, and Dabeka 2010) [2]
Soy-containing cereal mix (sold in Canada between 30/09 and 2/10/2008)	< 0.004 mg/kg	1	0	(Tittlemier, Lau, Menard, Corrigan, Sparling, Gaertner, Cao, and Dabeka 2010) [2]
Soy-containing drink mix (sold in Canada between 30/09 and 2/10/2008)	< 0.004 mg/kg	6	0	(Tittlemier, Lau, Menard, Corrigan, Sparling, Gaertner, Cao, and Dabeka 2010) [2]
Soy-containing yogurt (sold in Canada between 30/09 and 2/10/2008)	< 0.004 mg/kg	1	0	(Tittlemier, Lau, Menard, Corrigan, Sparling, Gaertner, Cao, and Dabeka 2010) [2]
Yogurt (sold in Canada between 30/09 and 2/10/2008)	< 0.004 mg/kg	23	0	(Tittlemier, Lau, Menard, Corrigan, Sparling, Gaertner, Cao, and Dabeka 2010) [2]

Table 1. (Continued)

Matrix	Melamine range	N. Samples	Positive Samples	References
Yogurt beverage (sold in Canada between 30/09 and 2/10/2008)	<0.004 - 0.00728 mg/kg	9	2	(Tittlemier, Lau, Menard, Corrigan, Sparling, Gaertner, Cao, and Dabeka 2010) [2]
Fruit juice milk for kids	< 0.20 mg/L	1	0	(Chen and Yan 2009)
Liquid milk	33.8 mg/L	1	1	(Chen and Yan 2009)
Powdered milk	< 0.25 mg/kg	1	0	(Chen and Yan 2009)
Powdered infant milk	1.32 - 23.63 mg/kg	5	3	(Chen and Yan 2009)
Powdered soy milk	31.73 mg/kg	1	1	(Chen and Yan 2009)
Liquid milk	0.28 mg/kg	NR	NR	(Miao et al. 2009) [3]
Milk tea	11.07 mg/kg	NR	NR	(Miao et al. 2009) [3]
Toffee	< 0.005 mg/kg	NR	NR	(Miao et al. 2009) [3]
Raw liquid milk	2.09 - 2.19 mg/kg	2	2	(Sun et al. 2010)
Semi-finished liquid milk	1.67 - 1.89 mg/kg	3	3	(Sun et al. 2010)
Flavored liquid milk	0.08 - 1.74 mg/kg	5	5	(Sun et al. 2010)
Ice cream, liquid milk, powdered milk	< 0.01 - 6175 mg/kg	105	47	(Xu et al. 2009)
Pork bun	< 0.5 mg/kg	NR	NR	(Fujita et al. 2009)
Kakuni bun	< 0.5 mg/kg	NR	NR	(Fujita et al. 2009)
Japanese tea bun with red bean and vanilla cream	4.0 mg/kg	1	1	(Fujita et al. 2009)
Cream bun	0.8 - 37.0 mg/kg	4	4	(Fujita et al. 2009)
Creamed corn crepe	13.6 mg/kg	1	1	(Fujita et al. 2009)
Milk powder and infant formula (sold in Africa between October and December 2008)	0.5 - 5.5 mg/kg	49	3	(Schoder 2010)
Egg-containing food items (sold in Canada between 3/11/2008 and 20/01/2009)	< 0.004 - 0.247 mg/kg	NR	NR	(Tittlemier, Lau, Menard, Corrigan, Sparling, Gaertner, Cao, Dabeka, et al. 2010) [2]
Whole eggs collected in Chinese market	< 0.01-0.206 mg/kg	42	6	(Xia et al. 2009)
Infant formula collected from major retail stores in Albany (New York) in 2008	0.00010 - 0.052 mg/kg	26	26	(Zhu and Kannan 2018) [4]

Matrix	Melamine range	N. Samples	Positive Samples	References
Soy-based meat substitutes and other soy-containing food items (sold in Canada between 3/11/2008 and 20/01/2009)	<0.004 - 0.0479 mg/kg	NR	NR	(Tittlemier, Lau, Menard, Corrigan, Sparling, Gaertner, Cao, Dabeka, et al. 2010) [2]
Fish and shrimp products (sold in Canada between 3/11/2008 and 20/01/2009)	<0.004 - 1.104 mg/kg	64	32	(Tittlemier, Lau, Menard, Corrigan, Sparling, Gaertner, Cao, Dabeka, et al. 2010) [2]
Vegetable products (sold in Canada between 3/11/2008 and 20/01/2009)	<0.004 - 0.688 mg/kg	100	46	(Tittlemier, Lau, Menard, Corrigan, Sparling, Gaertner, Cao, Dabeka, et al. 2010) [2]
UHT milk (sold in Turkey in June 2010)	<0.320 mg/kg	50	0	(Filazi et al. 2012)
Pasteurized milk (sold in Turkey in June 2010)	<0.200 mg/kg	50	0	(Filazi et al. 2012)
Powdered infant formula (sold in Turkey in June 2010)	<0.280 mg/kg	50	0	(Filazi et al. 2012)
Cheese (sold in Turkey in June 2010)	0.121 mg/kg	50	1	(Filazi et al. 2012)
Milk powder (sold in Turkey in June 2010)	0.505 - 0.86 mg/kg	50	4	(Filazi et al. 2012)
Yogurt (sold in Turkey in June 2010)	0.136 - 0.479 mg/kg	50	22	(Filazi et al. 2012)
Human breast milk (collected in Turkey between June and September 2010)	41.55 - 76.43 ng/L	77	16	(Yurdakok et al. 2014)
Human breast milk (collected in United States between 2009 and 2012)	<67 - 7140 ng/L	100	94	(Zhu and Kannan 2019b)
Milk powder (sold in Iranian market between January and September 2011)	1.50 - 30.32 mg/kg	9	9	(Hassani et al. 2013)
Liquid milk (sold in Iranian market between January and September 2011)	0.11 - 1.48 µg/mL	5	5	(Hassani et al. 2013)
Human-grade wheat gluten	<1 ng/mL	7	0	(Levinson and Gilbride 2011)
Human-grade rice protein concentrate	<1 - 16.2 ng/mL	7	3	(Levinson and Gilbride 2011)
Human-grade soy protein isolate	<1 ng/mL	7	0	(Levinson and Gilbride 2011)
Chicken eggs (collected in different traditional Taiwanese markets)	<0.025 mg/kg	8	0	(Wang et al. 2012)

NR = not reported.

[1] The reported value is a mean value.
[2] Food items analyzed as purchased.
[3] Not specified if the value is a mean, maximum or minimum.
[4] Food items analyzed as consumed.

Table 2. Ammelide, Ammeline, and Cyanuric Acid ranges in foods

Matrix	N. samples	Ammelide positive	Ammeline positive	Cyanuric Acid positive		References
Chinese infant fromulae (sold in Gansu province in 2008)	52	2.9 mg/kg	1.7 mg/kg	1.6 mg/kg		(Wu, Zhao, and Li 2009)[1]
Milk tea		< 0.005 mg/kg	< 0.005 mg/kg	< 0.005 mg/kg		(Miao et al. 2009)[1]
Fresh milk		0.073 mg/kg	< 0.005 mg/kg	1.64 mg/kg		(Miao et al. 2009)[1]
Toffee		< 0.005 mg/kg	< 0.005 mg/kg	0.86 mg/kg		(Miao et al. 2009)[1]
Infant formulae (sold in Canada in Autumn 2008)	94	0.0027 mg/kg	0.0009 mg/kg	0.080 mg/kg	89	(Braekevelt et al. 2011)
Human breast milk (sold in United States between 2009 and 2012)	100	< 0.02-0.77 ng/mL	< 0.027 - 1.19 ng/mL	0.033 - 6.26 ng/mL	69	(Zhu and Kannan 2019b)
Milk and milk powder (sold in Iran between Jan and Sept 2011)	14	NA	NA	< 1.2 mg/mL		(Hassani et al. 2013)
Human-grade wheat gluten	7	NA	NA	< 1 ng/mL		(Levinson and Gilbride 2011)
Human-grade rice protein concentrate	7	NA	NA	<1 - 5400 ng/mL		(Levinson and Gilbride 2011)
Human-grade soy protein isolate	7	NA	NA	< 1 ng/mL		(Levinson and Gilbride 2011)
Milk, milk powder and yogurt (collected in Chinese local market)	30	NA	NA	< 0.01 - 1.7 mg/kg		(Pan et al. 2013)
Infant formula (provided by Baoding Center for Disease Control and Prevention, China)	8	< 0.0165 mg/kg	< 0.010 mg/kg	< 0.0664 mg/kg		(Meng et al. 2015)
Infant formula sold in Albany (NY) in 2018	26	0.03 - 0.34 µg/kg	< 0.072 - 0.76 µg/kg	0.03 - 6.8 µg/kg	26	(Zhu and Kannan 2018)

Matrix	N. samples[1]	Ammelide positive	Ammeline positive	Cyanuric Acid positive		References
Milk sold in Albany (NY) between June and July 2018	10	0.18 - 5.0 µg/kg	0.08 - 0.27 µg/kg	10	0.30 - 3.2 µg/kg	10	(Zhu and Kannan 2018)
Yogurt sold in Albany (NY) between June and July 2018	8	0.33 - 0.83 µg/kg	0.08 - 0.27 µg/kg	8	<0.063 - 6.2 µg/kg	7	(Zhu and Kannan 2018)
Cheese sold in Albany (NY) between June and July 2018	8	0.43 - 5.4 µg/kg	0.16 - 0.41 µg/kg	8	2.2 – 38 µg/kg	8	(Zhu and Kannan 2018)
Butter sold in Albany (NY) between June and July 2018	8	0.03 - 3.17 µg/kg	0.04 - 0.16 µg/kg	8	<0.063 - 53 µg/kg	6	(Zhu and Kannan 2018)
Bread sold in Albany (NY) between June and July 2018	8	0.03 - 0.64 µg/kg	<0.072 - 2.5 µg/kg	7	0.06 – 9.7 µg/kg	8	(Zhu and Kannan 2018)
Meat sold in Albany (NY) between Aug and Sept 2018	28	<0.10 -4.50 µg/kg	<0.20 - 5.23 µg/kg	6	<0.25 - 303 µg/kg	21	(Zhu and Kannan 2019a)
Fish/Seafood sold in Albany (NY) between August and September 2018	11	<0.10 -7.16 µg/kg	<0.20 - 1.47 µg/kg	1	<0.25 - 206 µg/kg	5	(Zhu and Kannan 2019a)
Cereal products sold in Albany (NY) between August and September 2018	27	0.39 - 7.19 µg/kg	<0.20 - 8.875 µg/kg	22	<0.25 - 72.9 µg/kg	26	(Zhu and Kannan 2019a)
Vegetables sold in Albany (NY) between August and September 2018	13	0.17 - 4.20 µg/kg	<0.20 - 1.90 µg/kg	3	<0.25 - 308 µg/kg	4	(Zhu and Kannan 2019a)
Cooking oil sold in Albany (NY) between August and September 2018	10	<0.10 - 0.23 µg/kg	<0.20 - 43.10 µg/kg	9	<0.25 - 2.61 µg/kg	1	(Zhu and Kannan 2019a)
Beverages sold in Albany (NY) between August and September 2018	32	<0.10 - 1.90 µg/kg	<0.02 - 3.49 µg/kg	14	<0.03 - 28.4 µg/kg	31	(Zhu and Kannan 2019a)

NA = not analyzed.
[1] N. Samples not provided; not specified if the value is a mean, maximum or minimum.

AQSIQ also published data concerning melamine content in different food items collected in China in 2008, such as biscuits, cakes and confectionery, liquid milk and yoghurt, snack food, powdered milk and cereal products: melamine levels ranged between 0.09 and 6196.61 mg/kg (Table 1) (Hilts and Pelletier 2009), meaning that MEL contamination in China was not limited to infant formulae but it had spread to several food matrices.

In the same year Tittlemier, Lau, Menard, Corrigan, Sparling, Gaertner, Cao, and Dabeka (2010) analyzed a variety of dairy products and soy-based dairy substitutes (n = 208) sold in Canadian retail outlets between September 30 and October 2, 2008. Only 28 samples out of 208 food items showed MEL levels higher than the method quantitation limit (MQL = 0.004 mg/kg), which was more often present in food items containing milk (21%) than in soy-based dairy products substitutes (2%). Fortunately, aside from one candy sample, all MEL concentrations in the samples were at least 10 times lower than the maximum residual level of melamine allowed in foods (Table 1). These results were much lower than those reported by some authors in different Chinese dairy products (Chen and Yan 2009, Fujita et al. 2009, Miao et al. 2009, Sun et al. 2010, Xu et al. 2009) (Table 1). It is interesting to highlight that in the work of Chen and Yan (2009), among the different powered infant milk purchased from local supermarkets in Beijing in September 2008 no melamine was detected in samples imported from New Zealand and Netherlands, while the same Chinese food items contained MEL in the range (1.32-23.63) mg/kg.

In the same year, Schoder (2010) reported results regarding MEL content in different milk powders and infant formulae sold in East Africa between October and December 2008: only 6% (3 of 49) of all samples resulted positive to melamine, ranging between 0.5 and 5.5 mg/kg.

Later, Tittlemier, Lau, Menard, Corrigan, Sparling, Gaertner, Cao, Dabeka, et al. (2010) analyzed different food items (n = 364) sold in Canada in the period 3 November 2008–20 January 2009 for their melamine content. 26% of the samples were positive to MEL, and fish and vegetables products showed higher detection frequency than egg and soy products. Also in this case, all MEL concentrations were lower than 2.5 mg/kg allowed in foods (FAO 2010) (Table 1). The authors (Tittlemier, Lau, Menard, Corrigan,

Sparling, Gaertner, Cao, Dabeka, et al. 2010) attributed the presence of MEL in the vegetable products to the transformation of the pesticide CYRO to MEL, while in the case of fish products MEL was attributed to the incorporation of MEL into feed, since it has been shown that MEL bioaccumulates in the edible shrimp tissues (Andersen et al. 2008). The presence of MEL could also be attributed to the decomposition of trichloromelamine into MEL (WHO 2009), due to its use as sanitizer and disinfectant on hard food-contact surfaces and as a component of fruit and vegetable wash solutions in the USA (Tittlemier, Lau, Menard, Corrigan, Sparling, Gaertner, Cao, Dabeka, et al. 2010).

No or very low MEL contamination was also observed by Filazi et al. (2012) in different dairy products purchased from local groceries in Ankara (Turkey) in June 2010: among the 300 samples analyzed, only 27 resulted positive to melamine, with levels below 1 mg/kg (Table 1).

As infants are more vulnerable to melamine toxicity and breastfeeding could be cause of their melamine exposure, following mothers' contaminated foods consumption, Yurdakok et al. (2014) analyzed MEL levels in 77 breast milk obtained from lactating mothers in Ankara (Turkey) between June and September 2010: 16 samples resulted positive to melamine and only three breast milks had MEL content higher than the LOQ value (41.55 ng/L). Breast milk obtained from lactating mothers in the United States between 2009 and 2012 (Zhu and Kannan 2019b) showed much higher level of melamine concentration: 89 samples out of 100 had MEL levels ranging between 0.067 and 7.14 ng/mL, indicating the ongoing exposure of US lactating women to MEL. Furthermore, the same authors (Zhu and Kannan 2019b) also detected CYA, AMN and AMD in most of human milk samples analyzed (Table 2), cyanuric acid being the major compound, followed by melamine.

MEL contamination was also observed in milk and milk powder collected from the market in Tehran (Iran) in 2011: data reported by Hassani et al. (2013), in fact, showed MEL presence in all analyzed samples (see Table 1), with levels, in some cases, above the FAO limits (FAO 2010, 2012); no CYA was detected in the same samples (see Table 2).

Table 3. Continuing Occurrence of Melamine in foods: From 2012 to nowadays

Matrix	Melamine range	N. Samples	Positive Samples	References
Milk and milk powder (collected in Chinese local markets)	0.017 - 0.0349 mg/L	6	6	(Feng et al. 2012)
Milk, milk powder and yogurt (collected in Chinese local markets)	<0.02 - 0.62 mg/kg	30	7	(Pan et al. 2013)
Milk powder (sold in Uruguay during 2013 and 2014)	<0.019 - 0.082 mg/kg	40	9	(García Londoño et al. 2018)
Drink waters	110 - 240 ng/L	2	2	(Khedr 2013)[1]
Soft drinks	10 - 250 ng/L	5	5	(Khedr 2013)[1]
Dry milk infant formulae	0.0011 - 0.0024 mg/kg	4	4	(Khedr 2013)[1]
Fish (collected in Chinese local market)	<0.019 mg/kg	NR	NR	(Zhang, Ma, and Fan 2014)[1]
Shrimp (collected in Chinese local market)	<0.019 mg/kg	NR	NR	(Zhang, Ma, and Fan 2014)[1]
Winkle (collected in Chinese local market)	0.070 mg/kg	NR	NR	(Zhang, Ma, and Fan 2014)[1]
Clam (collected in Chinese local market)	0.189 mg/kg	NR	NR	(Zhang, Ma, and Fan 2014)[1]
Canned tuna (collected in Turkish supermarkets)	<0.68 mg/kg	8	0	(Demirhan et al. 2015)
Infant formula (provided by Baoding Center for Disease Control and Prevention, China)	<0.005 - 0.025 mg/kg	8	4	(Meng et al. 2015)
Domestic infant formula (collected in different Iranian drugstores)	<0.10 - 3.63 mg/kg	42	30	(Maleki et al. 2018)
Domestic follow up formula (collected in different Iranian drugstores)	<0.10 - 2.48 mg/kg	17	10	(Maleki et al. 2018)
Imported infant formula (collected in different Iranian drugstores)	<0.10 - 0.65 mg/kg	6	3	(Maleki et al. 2018)
Imported follow up formula (sold in Iran)	<0.10 - 0.38 mg/kg	4	2	(Maleki et al. 2018)
Poultry meat (sold in Iran during 2015-2016)	0.96 - 1.72 mg/kg	50	50	(Shakerian et al. 2018)
Hen eggs (sold in Iran during 2015-2016)	0.96 - 1.98 mg/kg	50	50	(Shakerian et al. 2018)
Ultra-filtration cheese (sold in Iran during 2015-2016)	0.45 - 1.62 mg/kg	50	50	(Shakerian et al. 2018)

Matrix	Melamine range	N. Samples	Positive Samples	References
Dairy cream (sold in Iran during 2015-2016)	0.81 - 2.18 mg/kg	50	50	(Shakerian et al. 2018)
Infant formula sold in Albany (NY) in 2018	0.00004 - 0.0027 mg/kg	26	26	(Zhu and Kannan 2018)[2]
Milk sold in Albany (NY) between June and July 2018	0.00084 - 0.0031 mg/kg	10	10	(Zhu and Kannan 2018)[2]
Yogurt sold in Albany (NY) between June and July 2018	0.00058 - 0.0019 mg/kg	8	8	(Zhu and Kannan 2018)[2]
Cheese sold in Albany (NY) between June and July 2018	0.00072 - 0.0037 mg/kg	8	8	(Zhu and Kannan 2018)[2]
Butter sold in Albany (NY) between June and July 2018	0.00048 - 0.0013 mg/kg	8	8	(Zhu and Kannan 2018)[2]
Bread sold in Albany (NY) between June and July 2018	0.00011 - 0.0028 mg/kg	8	8	(Zhu and Kannan 2018)[2]
Meat sold in Albany (NY) between Aug and Sept 2018	0.00034 - 0.00642 mg/kg	28	28	(Zhu and Kannan 2019a)
Fish/Seafood sold in Albany (NY) between Aug and Sept 2018	< 0.00012 - 0.00234 mg/kg	11	9	(Zhu and Kannan 2019a)
Cereal products sold in Albany (NY) between Aug and Sept 2018	0.00015 - 0.00933 mg/kg	27	27	(Zhu and Kannan 2019a)
Vegetables sold in Albany (NY) between Aug and Sept 2018	0.00023 - 0.00673 mg/kg	13	13	(Zhu and Kannan 2019a)
Cooking oil sold in Albany (NY) between Aug and Sept 2018	< 0.00012 - 0.00716 mg/kg	10	8	(Zhu and Kannan 2019a)
Beverages sold in Albany (NY) between Aug and Sept 2018	0.00005 - 0.0121 mg/kg	32	32	(Zhu and Kannan 2019a)

NR = not reported.
[1] Not specified if the value is a mean, maximum or minimum.
[2] Food items analyzed as consumed.

3.2. Continuing Occurrence of Melamine and Its Analogous in Foods: From 2012 to Nowadays

MEL contamination emergency seemed to be returned in China when Pan et al. (2013) analyzed different dairy products purchased from local market: results showed that among 30 products of milk, milk powder and yoghurt only 5 samples resulted positive to both MEL and CYA, and 2 additional samples were positive only to MEL, with levels in the range (0.06-1.7) mg/kg for CYA and (0.08-0.62) mg/kg for MEL.

Also according to Feng et al. (2012) MEL levels in dairy products collected in China were far below the maximum residual level of MEL allowed in foods (FAO 2010, FAO 2012) (see Table 3), and they were in the same order of magnitude of data reported by García Londoño et al. (2018) regarding MEL concentrations in milk powder purchased in Uruguayan supermarkets between 2013 and 2014 (see Table 3).

MEL contamination was not observed either in Turkey, where samples of canned tuna fish collected in different supermarkets in Ankara showed melamine levels much lower (Demirhan et al. 2015) than the Codex Alimentarius limit of 2.5 mg/kg allowed in food, even if the reported limit of quantification of the proposed method was not so low (LOQ = 0.68 mg/kg).

Zhu and Kannan (2018) compared MEL levels in some infant formulae purchased in different retail stores in Albany (New York) in 2018 with those obtained in some infant formulae purchased in Albany (New York) in 2008: all results were far below the maximum residual level of MEL allowed in foods (FAO 2010, 2012) (see Table 3). Furthermore, the sum of MEL and its related compounds significantly decreased from 2008 to 2018 (median values decreasing from 9.4 to 2.7 ng/g), which may reflect the effectiveness of the regulatory actions taken following MEL contamination incidents in 2007-2008. In the same work (Zhu and Kannan 2018), other food products (milk, yogurt, cheese, butter, and bread) collected in 2018 were analyzed for their levels of MEL and its related compounds: the target analytes were found in almost the samples analyzed (see Table 2 and Table 3), indicating the continuing occurrence of these compounds in food products, but, fortunately, all MEL concentrations were far below the maximum residual level set by FAO (FAO 2010). The authors (Zhu and Kannan 2018) also observed that CYA was the major compound found in dairy products in the US (68-83%), followed by MEL (7.0-21%), while ammelide and ammeline minimally contributed to the sum of MEL and its analogous.

Khedr (2013) investigated MEL content also in several marketed drink waters and soft drinks, all values being below 1 µg/L, while MEL levels in dry milk powders were far below 1 mg/kg: this is a further proof of the fact that melamine can be present in traces in several foods not due to contamination but to normal food production and processing.

In some countries there are still no regulations regarding MEL contamination in food products, and Iran is one of this (Maleki et al. 2018). That's why, still today, MEL contents in food products can result higher than the maximum residual level allowed in foods (FAO 2010, 2012). In the study of Maleki et al. (2018), for example, among several infant formulae (48) and follow up formulae (21) from different brands (5 domestic brands and 6 imported) randomly collected in various drugstores in Zanjan (Iran), 67.8% and 50% of domestic and imported samples, respectively, resulted positive to MEL: the maximum level was observed in a domestic infant formula (3.63 mg/kg). It's worthwhile noting that imported samples, probably coming from China (the major supplier of infant and follow up formula materials for Iran), had lower MEL contents compared to domestic food products (see Table 3), probably due the effectiveness of the regulatory actions taken in China and to the lack of regulation in Iran regarding MEL contamination in foods. MEL levels in the same order of magnitude were also found by Shakerian et al. (2018), who analyzed 200 food products (50 each of poultry meats, hen eggs, ultra-filtration cheeses, and dairy creams) collected from the major retailers in Iran during 2015-2016: all samples resulted positive to melamine, but all of them showed MEL contents below maximum residual level of 2.5 mg/kg.

In the recent years, Zhu and Kannan 2019 (Zhu and Kannan 2019a) provided a brief overview of MEL and its analogous levels in a range of food products marketed in the US (see Table 2 and Table 3): data reported can be considered as baseline levels of melamine and its related compounds in foodstuffs, with meat and cereal products showing the highest values of the sum of MEL and its analogous. Once again, data revealed the predominance of CYA in US foodstuff, followed by MEL (Zhu and Kannan 2018, Zhu and Kannan 2019a, b), which may reflect the contribution from other sources of CYA contamination, aside from MEL degradation, such as the use of CYA in animal feed or in water disinfection, or the fertilizers degradation.

CONCLUSION

Following the melamine food scandals occurred in 2007-2008, the attention focused on MEL prompted worldwide monitoring of its occurrence in different food matrices, as well as the development of several analytical techniques for its measurement. Since then, many analytical efforts have been made to detect MEL with an ever-increasing sensitivity and precision degree, so that in 2010 the International Organization for Standardization (ISO), in collaboration with the International Dairy Federation (IDF), published the technical specification for the quantitative determination of melamine and cyanuric acid by LC-MS/MS in milk, milk products and infant formulae. Nowadays, it is still the standard method for MEL and CYA determination.

Ten years after melamine adulteration incidents, the effectiveness of the regulatory actions taken in different countries resulted in a significant decrease of MEL content and its analogous in foods. Apart from poor food recalls, data reported in the literature showed the continuing occurrence of melamine and its derivatives throughout the food supply chain, but, fortunately, most of the current MEL levels are low and they do not constitute a health risk for adult consumers. Those countries having no regulations regarding melamine contamination in food products are an exception, and it would be desirable they adopt some regulations as soon as possible.

Even if most of the current MEL levels in food items are far below the maximum residual level of melamine allowed in foods, this chemical compound is widespread in a wide range of food products. Therefore, it is necessary to continuously monitor MEL residues throughout the food supply chain in order to prevent excessive human exposure to melamine *via* food consumption.

REFERENCES

Andersen, W. C., S. B. Turnipseed, C. M. Karbiwnyk, S. B. Clark, M. R. Madson, C. M. Gieseker, R. A. Miller, N. G. Rummel, and R. Reimschuessel. 2008. "Determination and Confirmation of Melamine Residues in Catfish, Trout, Tilapia, Salmon, and Shrimp by Liquid Chromatography with Tandem Mass Spectrometry." *Journal of Agricultural and Food Chemistry* 56:4340–47.

AOAC. 2005. *Official methods of analysis*. Edited by AOAC International. 17th ed. Gaithersburg, Md.

Arnold, E. 1990. "Cyromazine explanation." *IPCS INCHEM cooperation between the International Programme on Chemical Safety (IPCS) and the Canadian Centre for Occupational Health and Safety (CCOHS)*, accessed October, 2019. http://www.inchem.org/documents/jmpr/jmpmono/v90pr06.htm.

BIPM. 2012. *International vocabulary of metrology – Basic and general concepts and associated terms*. Edited by JCGM. VIM 3rd ed. Vol. JCGM 200:2012(E/F).

Braekevelt, E., B. P. Lau, S. Feng, C. Menard, and S. A. Tittlemier. 2011. "Determination of melamine, ammeline, ammelide and cyanuric acid in infant formula purchased in Canada by liquid chromatography-tandem mass spectrometry." *Food Additives and Contaminants* 28 (6):698-704. doi: 10.1080/19440049.2010.545442.

Chen, Z., and X. Yan. 2009. "Simultaneous determination of melamine and 5-hydroxymethylfurfural in milk by capillary electrophoresis with diode array detection." *Journal of Agricultural and Food Chemistry* 57 (19):8742-7. doi: 10.1021/jf9021916.

Commission of the EU Communities. 2002. "Commission Decision of 12 August 2002 implementing Council Directive 96/23/EC concerning the performance of analytical methods and the interpretation of results." Ed *Official Journal of the European Communities.*

Demirhan, B. E., B. Demirhan, S. Y. BAŞ, G. Yentür, and A. B. Öktem. 2015. "Investigation of Melamine Presence in Canned Tuna Fish by Capillary Zone Electrophoresis Method." *Journal of Agricultural Sciences* 21 (2):310-15.

Domingo, E., A. A. Tirelli, C. A. Nunes, M. C. Guerreiro, and S. M. Pinto. 2014. "Melamine detection in milk using vibrational spectroscopy and chemometrics analysis: A review." *Food Research International* 60:131-39. doi: 10.1016/j.foodres.2013.11.006.

Ehling, S., S. Tefera, and I. P. Ho. 2007. "High-performance liquid chromatographic method for the simultaneous detection of the adulteration of cereal flours with melamine and related triazine by-products ammeline, ammelide, and cyanuric acid." *Food Additives & Contaminants* 24 (12):1319-25. doi: 10.1080/02652030701673422.

Eurachem. 2014. "*Eurachem Guide: The Fitness for Purpose of Analytical Methods – A Laboratory Guide to Method Validation and Related Topics.*" ed B. Magnusson and U. Örnemark. https://www.eurachem.org/index.php/publications/guides/mv.

FAO. 2010. "*Food and Agriculture Organization of the United Nations, International Experts Limit Melamine Levels in Food.*" ed.

FAO. 2012. *Food and Agriculture Organization of the United Nations, Codex Committee on Contaminants in Foods - Draft maximum levels for melamine in food (liquid infant formula).*

Feng, W., C. Lv, L. Yang, J. Cheng, and C. Yan. 2012. "Determination of melamine concentrations in dairy samples." *LWT - Food Science and Technology* 47 (1):147-53. doi: 10.1016/j.lwt.2011.12.021.

Filazi, A., U. T. Sireli, H. Ekici, H. Y. Can, and A. Karagoz. 2012. "Determination of melamine in milk and dairy products by high performance liquid chromatography." *Journal of Dairy Science* 95 (2):602-8. doi: 10.3168/jds.2011-4926.

Food and Drug Administration. 2008. "*Food additives permitted in feed and drinking water of animals: feed-grade biuret*, Title 21, Volume 6,

Section 573. 220." accessed October 2019. https://www.accessdata. fda.gov/scripts/cdrh/cfdocs/cfCFR/CFRSearch.cfm?fr=573.220.

Food and Drug Administration. 2018. "*Bioanalytical Method Validation Guidance for Industry.*" ed USFDA Beltsville, MD. https://www.fda.gov/regulatory-information/search-fda-guidance-documents/bioanalytical-method-validation-guidance-industry.

Fujita, M., K. Kakimoto, Nagayoshi H, Konishi Y., Uchida K., Osakada M., M. Okihaschi, and H. Obana. 2009. "Determination of Melamine in Chinese-made Processed Food." *Shokuhin Eiseigaku Zasshi* 50 (3):131-34.

García Londoño, V. A., M. Puñales, M. Reynoso, and S. Resnik. 2018. "Melamine contamination in milk powder in Uruguay." *Food Addit Contam Part B Surveill* 11 (1):15-19. doi: 10.1080/19393210.2017. 1389993.

Ge, X., X. Wu, J. Wang, S. Liang, and H. Sun. 2015. "Highly sensitive determination of cyromazine, melamine, and their metabolites in milk by molecularly imprinted solid-phase extraction combined with ultra-performance liquid chromatography." *Journal of Dairy Science* 98 (4):2161-71. doi: 10.3168/jds.2014-8793.

Gossner, C. M., J. Schlundt, P. Ben Embarek, S. Hird, D. Lo-Fo-Wong, J. J. Beltran, K. N. Teoh, and A. Tritscher. 2009. "The melamine incident: implications for international food and feed safety." *Environmental Health Perspectives* 117 (12):1803-8. doi: 10.1289/ehp.0900949.

Gratz, S., B. Gamble, and D. Heitkemper. 2009. "Screen for the presence of melamine and cyanuric acid in milk-based and soy-based foods using LC-ESI-MSn." In *Laboratory Information Bulletins LIB No. 4427*, edited by U.S. Food and Drug Administration. https://www.fda.gov/food/laboratory-methods-food/laboratory-information-bulletin-lib-4422-melamine-and-cyanuric-acid-residues-foods.

Hassani, S., F. Tavakoli, M. Amini, F. Kobarfard, A. Nili-Ahmadabadi, and O. Sabzevari. 2013. "Occurrence of melamine contamination in powder and liquid milk in market of Iran." *Food Addit Contam Part A Chem Anal Control Expo Risk Assess* 30 (3):413-20. doi: 10.1080/19440049. 2012.761730.

Hilts, C., and L. Pelletier. 2009. "*Background Paper on Occurrence of Melamine in Foods and Fee*d." ed WHO Ottawa, Canada. https://www.who.int/foodsafety/fs_management/Melamine_3.pdf (accessed October 2019).

Ishiwata, H., T. Inoue, T. Yamazaki, and K. Yoshihira. 1987. "Liquid chromatographic determination of melamine in beverages." *Journal-Association of Official Analytical Chemists* 70 (3):457-60.

ISO-IDF. 2010. "*Milk, milk products and infant formulae - Guideline for the quantitative determination of melamine and cyanuric acid by LC-MS/MS.*" ed Geneva: Technical Committee ISO/TC 34, Food products, Subcommittee SC 5, and the International Dairy Federation (IDF).

Joint FAO/WHO Expert Committee on Food Additives. 2010. *Meeting, and World Health Organization. Evaluation of Certain Food Additives: Seventy-first Report of the Joint FAO/WHO Expert Committee on Food Additives.* Vol. 71: World Health Organization.

Karbiwnyk, C. M., W. C. Andersen, S. B. Turnipseed, J. M. Storey, M. R. Madson, K. E. Miller, C. M. Gieseker, R. A. Miller, N. G. Rummel, and R. Reimschuessel. 2009. "Determination of cyanuric acid residues in catfish, trout, tilapia, salmon and shrimp by liquid chromatography-tandem mass spectrometry." *Analytica Chimica Acta* 637 (1-2):101-11. doi: 10.1016/j.aca.2008.08.037.

Khedr, A. 2013. "Optimized extraction method for LC-MS determination of bisphenol A, melamine and di(2-ethylhexyl) phthalate in selected soft drinks, syringes, and milk powder." *Journal of Chromatography B* 930:98-103. doi: 10.1016/j.jchromb.2013.04.040.

Kowalsky, L. 1992. "Certified pool-SPA operator." *National Swimming Pool Fundation: Taxas* 46.

Levinson, L. R., and K. A. Gilbride. 2011. "Detection of Melamine and Cyanuric Acid in Vegetable Protein Products Used in Food Production." *Journal of Food Science* 76 (4):C568-C75. doi: 10.1111/j.1750-3841.2011.02148.x.

Litzau, J. J., G. E. Mercer, and K. J. Mulligan. 2008. "GC-MS Screen for the Presence of Melamine, Ammeline, Ammelide, and Cyanuric Acid." In *Laboratory Information Bulletin* LIB No. 4423, edited by U.S. Food and

Drug Administration. https://www.fda.gov/food/laboratory-methods-food/laboratory-information-bulletin-lib-4423-melamine-and-related-compounds.

Maleki, J., F. Nazari, J. Yousefi, R. Khosrokhavar, and M. J. Hosseini. 2018. "Determinations of Melamine Residue in Infant Formula Brands Available in Iran Market Using by HPLC Method." *Iranian Journal of Pharmaceutical Research* 17 (2):563-70.

Meng, Z., Z. Shi, S. Liang, X. Dong, Y. Lv, and H. Sun. 2015. "Rapid screening and quantification of cyromazine, melamine, ammelide, ammeline, cyanuric acid, and dicyandiamide in infant formula by ultra-performance liquid chromatography coupled with quadrupole time-of-flight mass spectrometry and triple quadrupole mass spectrometry." *Food Control* 55:158-65. doi: 10.1016/j.foodcont.2015.02.034.

Miao, H., S. Fan, Y. N. Wu, L. Zhang, P. P. Zhou, J. G. Li, H. J. Chen, and Y. F. Zhao. 2009. "Simultaneous Determination of Melamine, Ammelide, Ammeline, and Cyanuric Acid in Milk and Milk Products by Gas Chromatography-tandem Mass Spectrometry." *Biomedical and Environmental Sciences* 22:87-94.

Moore, Jeffrey C., Jonathan W. DeVries, Markus Lipp, James C. Griffiths, and Darrell R. Abernethy. 2010. "Total Protein Methods and Their Potential Utility to Reduce the Risk of Food Protein Adulteration." *Comprehensive Reviews in Food Science and Food Safety* 9 (4):330-57. doi: 10.1111/j.1541-4337.2010.00114.x.

Nascimento, C. F., P. M. Santos, E. R. Pereira-Filho, and F. R. P. Rocha. 2017. "Recent advances on determination of milk adulterants." *Food Chemistry* 221:1232-44. doi: 10.1016/j.foodchem.2016.11.034.

Pan, X.-D., P.-g Wu, D.-J. Yang, L.-Y. Wang, X.-H. Shen, and C.-Y. Zhu. 2013. "Simultaneous determination of melamine and cyanuric acid in dairy products by mixed-mode solid phase extraction and GC–MS." *Food Control* 30 (2):545-48. doi: 10.1016/j.foodcont.2012.06.045.

Pichon, V., L. Chen, S. Guenu, and M. C. Hennion. 1995. "Comparison of sorbents for the solid-phase extraction of the highly polar degradation products of atrazine (including ammeline, ammelide and cyanuric acid)." *Journal of Chromatography A* 711 (2):257-67.

Ritota, M., and P. Manzi. 2017. "Melamine Detection in Milk and Dairy Products: Traditional Analytical Methods and Recent Developments." *Food Analytical Methods* 11 (1):128-47. doi: 10.1007/s12161-017-0984-1.

Root, D. S., T. Hongtrakult, and W. C. Dauterman. 1996. "Studies on the AbsorDtion. Residues and Metabolism of Cyromazine in Tomatoes." *Pesticide Science* 48:25-30.

Rovina, Kobun, and Shafiquzzaman Siddiquee. 2015. "A review of recent advances in melamine detection techniques." *Journal of Food Composition and Analysis* 43:25-38. doi: 10.1016/j.jfca.2015.04.008.

Sancho, J. V., M. Ibáñez, S. Grimalt, Ó J. Pozo, and F. Hernández. 2005. "Residue determination of cyromazine and its metabolite melamine in chard samples by ion-pair liquid chromatography coupled to electrospray tandem mass spectrometry." *Analytica Chimica Acta* 530 (2):237-43. doi: https://doi.org/10.1016/j.aca.2004.09.038.

Schoder, D. 2010. "Melamine milk powder and infant formula sold in East Africa." *Journal of Food Protection* 73 (9):1709-14. doi: 10.4315/0362-028x-73.9.1709.

Shakerian, A., F. Khamesipour, E. Rahimi, P. Kiani, M. M. Shahraki, M. Pooyan, S. Hemmatzadeh, and Y. C. Tyan. 2018. "Melamine levels in food products of animal origin in Iran." *Revue de Medecine Veterinaire* 169 (7-9):152-56.

Smoker, M., and A. J. Krynitsky. 2008. "Interim Method for Determination of Melamine and Cyanuric Acid Residues in Foods using LC-MS/MS: Version 1.0." In *Laboratory Information Bulletin* LIB No. 4422, edited by U.S. Food and Drug Administration. https://www.fda.gov/food/laboratory-methods-food/laboratory-information-bulletin-lib-4422-melamine-and-cyanuric-acid-residues-foods.

Sun, H., L. Wang, L. Ai, S. Liang, and H. Wu. 2010. "A sensitive and validated method for determination of melamine residue in liquid milk by reversed phase high-performance liquid chromatography with solid-phase extraction." *Food Control* 21 (5):686-91. doi: 10.1016/j.foodcont.2009.10.008.

Tittlemier, S. A. 2010. "Methods for the analysis of melamine and related compounds in foods: a review." *Food Additives and Contaminants* 27 (2):129-45. doi: 10.1080/19440040903289720.

Tittlemier, S. A., B. P. Lau, C. Menard, C. Corrigan, M. Sparling, D. Gaertner, X. L. Cao, and B. Dabeka. 2010. "Baseline levels of melamine in food items sold in Canada. I. Dairy products and soy-based dairy replacement products." *Food Additives and Contaminants* 3 (3):135-9. doi: 10.1080/19440049.2010.502654.

Tittlemier, S. A., B. P. Lau, C. Menard, C. Corrigan, M. Sparling, D. Gaertner, X. L. Cao, B. Dabeka, and C. Hilts. 2010. "Baseline levels of melamine in food items sold in Canada. II. Egg, soy, vegetable, fish and shrimp products." *Food Additives and Contaminants* 3 (3):140-7. doi: 10.1080/19440049.2010.502655.

Tittlemier, S. A., B. P. Lau, C. Menard, C. Corrigan, M. Sparling, D. Gaertner, K. Pepper, and M. Feeley. 2009. "Melamine in infant formula sold in Canada: occurrence and risk assessment." *Journal of Agricultural and Food Chemistry* 57 (12):5340-4. doi: 10.1021/jf9005609.

Valat, C., P. Marchand, B. Veyrand, M. Amelot, C. Burel, N. Eterradossi, and G. Postollec. 2011. "Transfer of melamine in some poultry products." *Poultry Science* 90 (6):1358-63. doi: 10.3382/ps.2010-01205.

Vallejo-Cordoba, B., and A. F. Gonzalez-Cordova. 2010. "Capillary electrophoresis for the analysis of contaminants in emerging food safety issues and food traceability." *Electrophoresis* 31 (13):2154-64. doi: 10.1002/elps.200900777.

Varelis, P., and R. Jeskelis. 2008. "Preparation of [13C3]-melamine and [13C3]-cyanuric acid and their application to the analysis of melamine and cyanuric acid in meat and pet food using liquid chromatography-tandem mass spectrometry." *Food Additives and Contaminants* 25 (10):1208-15. doi: 10.1080/02652030802101893.

Wang, P. C., R. J. Lee, C. Y. Chen, C. C. Chou, and M. R. Lee. 2012. "Determination of cyromazine and melamine in chicken eggs using quick, easy, cheap, effective, rugged and safe (QuEChERS) extraction coupled with liquid chromatography-tandem mass spectrometry." *Analytica Chimica Acta* 752:78-86. doi: 10.1016/j.aca.2012.09.029.

Wang, T., J. Ma, Y. Chen, Y. Li, L. Zhang, and Y. Zhang. 2016. "Analysis of melamine and analogs in complex matrices: Advances and trends." *Journal of Separation Science* 40 (1):170-82. doi: 10.1002/jssc.201600854.

WHO. 2009. *"Toxicological and Health Aspects of Melamine and Cyanuric Acid: Report of a WHO Expert Meeting In collaboration with FAO Supported by Health Canada, 1-4 December 2008."* ed WHO Ottawa, Canada. https://www.who.int/foodsafety/publications/chem/Melamine_report09.pdf (accessed October 2019).

Wu, Yong-Ning, Yun-Feng Zhao, and Jin-Guang Li. 2009. "A Survey on Occurrence of Melamine and Its Analogues in Tainted Infant Formula in China." *Biomedical and Environmental Sciences* 22 (2):95-99. doi: https://doi.org/10.1016/S0895-3988(09)60028-3.

Xia, J., N. Zhou, Y. Liu, B. Chen, Y. Wu, and S. Yao. 2010. "Simultaneous determination of melamine and related compounds by capillary zone electrophoresis." *Food Control* 21 (6):912-18. doi: https://doi.org/10.1016/j.foodcont.2009.12.009.

Xia, X., S. Ding, X. Li, X. Gong, S. Zhang, H. Jiang, J. Li, and J. Shen. 2009. "Validation of a confirmatory method for the determination of melamine in egg by gas chromatography-mass spectrometry and ultra-performance liquid chromatography-tandem mass spectrometry." *Analytica Chimica Acta* 651 (2):196-200. doi: 10.1016/j.aca.2009.08.025.

Xu, X. M., Y. P. Ren, Y. Zhu, Z. X. Cai, J. L. Han, B. F. Huang, and Y. Zhu. 2009. "Direct determination of melamine in dairy products by gas chromatography/mass spectrometry with coupled column separation." *Analytica Chimica Acta* 650 (1):39-43. doi: 10.1016/j.aca.2009.04.026.

Yurdakok, B., A. Filazi, H. Ekici, T. H. Celik, and U. T. Sireli. 2014. "Melamine in breast milk." *Toxicology Research* 3 (4):242-46. doi: 10.1039/c3tx50095k.

Zhang, Y., X. Ma, and Y. Fan. 2014. "A Rapid and Sensitive Method for Determination of Melamine in Fish, Shrimp, Clam, and Winkle by Gas

Chromatography–Mass Spectrometry with Microwave-Assisted Derivatization." *Food Analytical Methods* 7 (9):1763-69. doi: 10.1007/s12161-014-9810-1.

Zhu, H., and K. Kannan. 2019a. "Melamine and cyanuric acid in foodstuffs from the United States and their implications for human exposure." *Environment International* 130:104950. doi: 10.1016/j.envint.2019.104950.

Zhu, H., and K. Kannan. 2019b. "Occurrence of Melamine and Its Derivatives in Breast Milk from the United States and Its Implications for Exposure in Infants." *Environmental Science & Technology* 53 (13):7859-65. doi: 10.1021/acs.est.9b02040.

Zhu, Hongkai, and Kurunthachalam Kannan. 2018. "Continuing Occurrence of Melamine and Its Derivatives in Infant Formula and Dairy Products from the United States: Implications for Environmental Sources." *Environmental Science & Technology Letters* 5 (11):641-48. doi: 10.1021/acs.estlett.8b00515.

In: An Introduction to Melamine
Editor: Ashley Harris
ISBN: 978-1-53617-136-5
© 2020 Nova Science Publishers, Inc.

Chapter 2

INFLUENCE ONTO THE IMMATURES: MELAMINE FROM MOTHERS TO BABIES

Ching Yan Chu, PhD and Chi Chiu Wang[*], *MD, PhD*

Department of Obstetrics and Gynaecology,
The Chinese University of Hong Kong, Shatin, Hong Kong
Li Ka Shing Institute of Health Sciences,
The Chinese University of Hong Kong, Shatin, Hong Kong
School of Biomedical Sciences,
The Chinese University of Hong Kong, Shatin, Hong Kong

ABSTRACT

In the last 2 decades, melamine contamination of animal feed and infant formulae has raised a global concern on food safety. Since then researches on the compound manufactured for the past 50 years have been re-activated. Most of the studies mainly focused on renal toxicity. Timely clinical treatments to the affected babies of the scandal have been convincingly beneficial. However, studies of reproductive and developmental toxicity of melamine on pregnant mothers, developing

---

[*] Corresponding Author's Email: ccwang@cuhk.edu.hk.

foetuses and neonates have been lacking. Although high level of adulteration has ceased, melamine is still a popular material for pesticides, farm animal feed fillers, fire retardants, anti-wrinkles and mild abrasives, etc. Low-dose contamination of melamine to the environment cannot be ignored. Animal models demonstrated the low-dose melamine transfer from mothers to foetuses and/or neonates, toxicokinetics in the pregnant, foetal and neonatal rats and reproductive and developmental toxicology at different stages of development. Detailed studies on the long-term effects on the immatures are still very limited.

Although there have been no clinical reports of this kind of data, animal experiments have proven the transfer of melamine from the mothers to their offspring, with evidence in birds and mammals in particular. Presence of melamine in eggs, muscles and various organs of the offspring has been reported. Loss of embryos, delay of foetal development, morphological changes in the kidneys and bone forming centres have also been revealed. On the other hand, it is also worth-considering the change of the maternal physiology during pregnancy that ingestion of melamine in daily diet or the contaminated maternal formulae can post risks that are different from those to the non-pregnant individuals. As it is well known that kidneys can compromise during pregnancy, an extra stress exerted by melamine may further complicate the condition. With these evidences, we can postulate that acute kidney injury may not be the whole story of melamine effects on the infants.

1. INTRODUCTION

Melamine, 1,3,5-Triazine-2,4,6-triamine, rose to fame after the pet food scandal in the North America in 2007 and drew maximum attention in 2008 when more than 300,000 Chinese infants were affected by the melamine-containing baby formulae. However, the story commenced as early as in 2003 that pets in South Africa, South Korea and Taiwan suffered the same nephrotoxicity as seen in pets later in the same decade. The most noticeable difference between these incidents is that the affected pets were mostly adults while the concerned humans were mainly children, and even infants. Before the melamine infant formula scandal, there were only very limited studies on the safety upon melamine ingestion. Since then, the majority of the studies focused on the farm animal or lab animal adults. Data collected from infants were largely from the clinical observations in China.

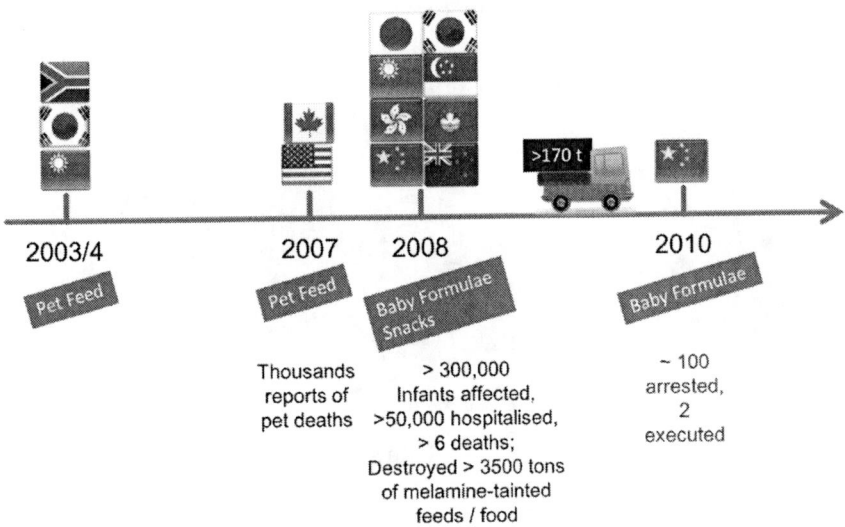

Figure 1. The melamine scandal timeline.

Melamine is everywhere, before and after the melamine scandals, and added to food both unintentionally and deliberately. Melamine was found in wheat gluten, rice protein, corn gluten, baby formulae as well as snacks using dairy products as ingredients. In order to harmonise the international food standards to protect the health of consumers and ensure fair trade practices, the Codex Alimentarius Commission has set a collection of food standards. In the case of an infant formula, the Codex Alimentarius Commission has stated that the prepared formula ready for consumption shall contain 1.8-3.0 g protein/100 kcal (Codex Alimentarius Commission 2016). Therefore, the industrial synthesised, nitrogen-rich melamine has been purposefully added to animal feed and cow's milk for human consumption so that it can produce a fraudulent result of Kjeldahl test that can reduce very much the cost of production of 'protein.' Although there is a review of melamine inefficiently replacing urea as a non-protein nitrogen source for livestock due to incomplete hydrolysis (Newton and Utley 1978), melamine is still constantly added to animal feed as filler, for both the ruminants and the monogastrics (New York Times 2007). In 2008, the General Administration of Quality Supervision, Inspection and Quarantine (AQSIQ) of China released a report that every kg of the infant formulae in the Chinese market contained 0.09–619 mg melamine (China Daily 2008). A scientific report, on the other hand, stated that the highest melamine content measured in the infant formulae was 4700 mg/kg (Jia, et al. 2009).

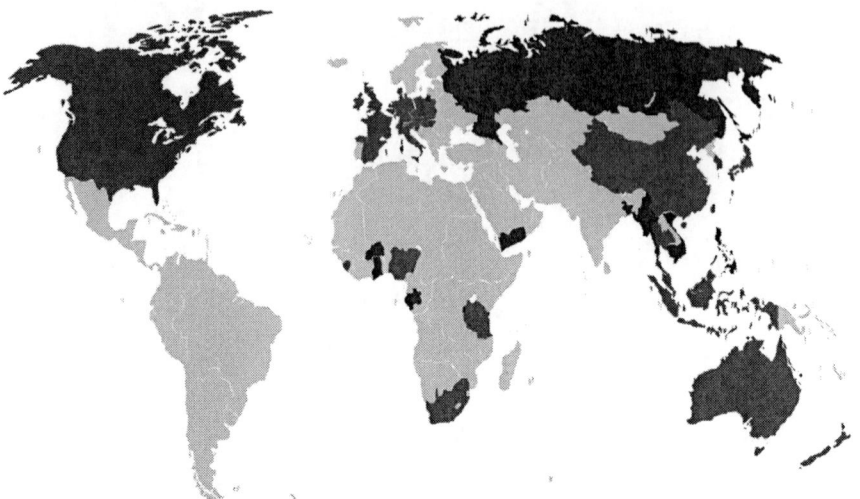

■ Countries with melamine findings in products or with ingredients originated from China. The positive results were transmitted to WHO.
■ Countries reported the import of the contaminated products, or the reports were made by the exporting country.

Figure 2. Map of the affected countries. Global distribution of melamine-contaminated products as reported to INFOSAN and published on national official websites. Adopted from Grossner, *et al*, The Melamine Incident: Implications for International Food and Feed Safety, Environ Health Perspect. 2009 Dec; 117(12): 1803–1808.

Since the food fraud in 2008, many countries have issued an alert on the presence of melamine in food; therefore, brazen high-dose adulteration has eventually ceased. Unfortunately, melamine is still being used in many areas that is not directly related to food at all. It is used as kitchenware, coating of food-contact containers, plastic packages, melamine resin, cleansing foam, paints, fire retardants, laminates for tabletops, adhesives, textiles, wastewater treatment systems etc., for its durability, fire-retarding, stain-repelling, water-repelling and shrink-resisting properties. Aging and mechanical wearing of these materials are probably releasing melamine to the environment. Moreover, triazine, the group that melamine belongs to

(Jutzi, Cook and Hütter 1982) (Yokley, et al. 2000) (Wackett, et al. 2002), is often used as herbicides, while cyromazine is a known precursor of melamine (Lim, et al. 1990) (Root, Hongtrakul and Dauterman 1996) (Yokley, et al. 2000) (Wackett, et al. 2002) (Wang, et al. 2014). Addition of these pesticides can cause the contamination of the farmland and the produce by melamine.

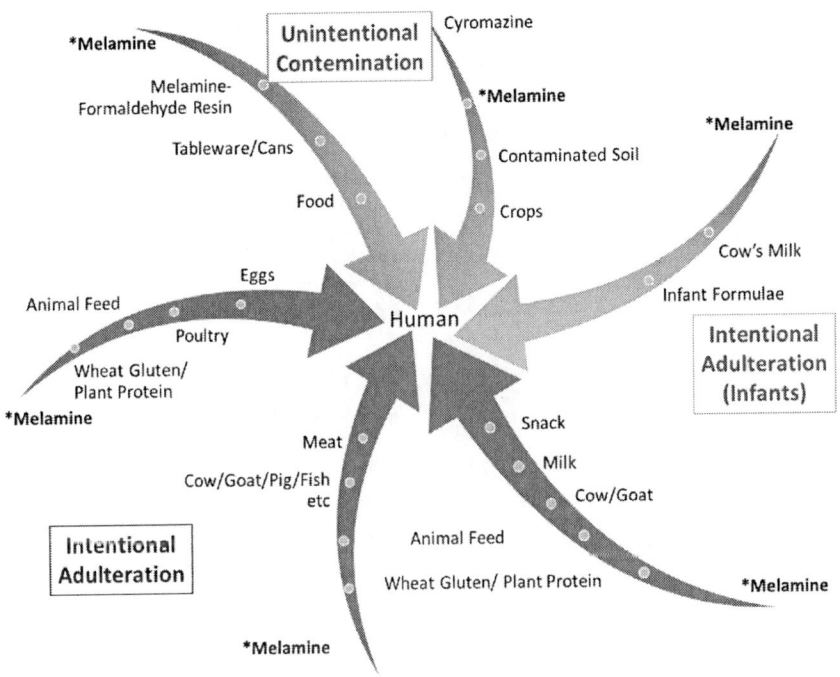

Figure 3. Human exposure to melamine.

The common diagnosis of the affected-farm animals, family pets and the human infants were acute renal injury (AKI, previously known as acute renal failure). The clinical presentations of the family pets were inappetence, vomiting, polyuria, polydipsia, lethargy, azotemia and hyperphosphatemia (Brown, et al. 2007) (Cianciolo, et al. 2008). The diagnosis of AKI was confirmed by pathology of other studies, ranging from distal tubular necrosis, intra-tubular crystals, chronic tubulointerstitial nephritis, mild anisokaryosis, focal proliferation of tubular epithelial cells, renal interstitial fibrosis, lymphoplasmacytic inflammation (Jeong, et al. 2006) (Brown, et al. 2007), amorphous, rounded, fan-shaped crystals in urine (Puschner, et al. 2007), intratubular, round, yellow-brown crystals with radiating striations, dilation of the proximal and the distal tubules (Nilubol, et al. 2009), severe interstitial nephritis and fibrosis with crystal deposition (González, et al. 2009).

Clinical observations of the infants affected by the melamine-laced formulae also suggested paediatric AKI. Complaints included crying on urination, discontinuance during urination, frequent diarrhea and fever, haematuria and hydronephrosis. Many of them also suffered from nephrolithiasis and urolithiasis, and required hospitalisation (Sun, et al. 2009) (Zhang, et al. 2009) (Yang, et al. 2010) (Yang, et al. 2010) (Wen, et al. 2011) (Shang, et al. 2012) (Shi, et al. 2012) (Hu, Wang and Hu, et al. 2013). The incidence of the urinary tract calculi was found to have positive correlation to the concentration of melamine in the milk formulae (Zhang, et al. 2009) (Shi, et al. 2012).

The co-contamination by cyanuric acid, which can form a lower-solubility complex with melamine (Ma and Bong 2011), made the injury to the kidney worse. Being a byproduct of the melamine synthesis, the existence of cyanuric acid was not unexpected. Although melamine metabolism in mammalian cells is insignificant (Panesar, et al. 2010), the gut microflora has the ability to turn melamine into its derivatives. In well-controlled mice experiments, administration of solely melamine can still result in detection of a minute amount of cyanuric acid in the feces. Contrariwise, application of antibiotics to these mice can suppress the conversion of melamine into its metabolites. In that experiment, *Klebsiella terrigena*, *Klebsiella pneumonia* and *Klebsiella planticola* were identified as the bacteria responsible for the conversion (Zheng, et al. 2013), which also reside in the guts of human and pig (Conlan, Kong and Segre 2012) (Dai, et al. 2012). Other bacteria were also found to be able to metabolise melamine, e.g., *Pseudomonas* sp (Jutzi, Cook and Hütter 1982) (Eaton and Karns 1991) (Cheng, et al. 2005), *Nocardioides* sp (Takagi, et al. 2012), *Escherichia coli* K-12 MG1655 (Cheng, et al. 2005), *Micrococcus* sp MF-1 (El-Sayed, El-Baz and Othman 2006) and *Rhodococcus* sp strain Mel

(Dodge, Wackett and Sadowsky 2012). Hence, many of the researchers has included also cyanuric acid in their studies. In spite of that, analysis of the melamine-associated urinary stones provided evidence of melamine co-crystallising with uric acid (Chang, Shi, et al. 2012), making the existence of cyanuric acid not essential to the urinary stone formation.

Infants are vulnerable; their immature systems cannot handle xenobiotics as well as the adults can. Moreover, infantile or juvenile diseases may give rise to health problems in later life. There are 2 periods of time for this vulnerable group to contact the xenobiotic: 1. *in utero* i.e., before birth; 2. *ex utero* i.e., after birth.

2. IN UTERO CONTACT OF MELAMINE

2.1. Placental Transfer of Melamine

The only source of nutrition of a developing foetus is brought through the placenta from its own mother. As long as a molecule can pass through the placenta, it can be delivered to the foetus. There are many pieces of evidence showing that melamine reaches various organs in adult animals, including rodents, ruminants, birds and fish (Melnick, et al. 1984) (Ogasawara, et al. 1995) (Chen, et al. 2009) (Chu, Chu and Chan, et al. 2010) (Gao, et al. 2010) (Reimschuessel, et al. 2010) (P. Sun, J. Wang, et al., Residues of melamine and cyanuric acid in milk and tissues of dairy cows fed different doses of melamine 2011) (Chan, et al. 2011) (Zhang, Guo and Wang 2012) (Chu, Chu and Ho, et al. 2013) (H. Sun, et al. 2016) although according to a study using 14C labelled melamine, the majority, 90%, of the administered melamine is excreted *per se* in urine in 24 hours (Mast, et al. 1983). Evidence on the placental transfer proves that melamine contact can be as early as in the foetal stage.

Direct and indirect evidence of placental transfer were provided not long after the 2008 incidents (Jingbin, et al. 2010) (Chu, Chu and Chan, et al. 2010) (Chan, et al. 2011) (Kim, Lee and Lim, et al. 2011) (Partanen, et al. 2012) (Chu, Chu and Ho, et al. 2013) (Kim, Lee and Baek, et al. 2013) (Chu,

Tang, et al. 2017). Partanen's group has shown 34–45% of the melamine added can be perfused through human term placenta in 5 minutes, but no crystal, significant oedema or syntiotrophoblastic vacuolisation was observed. On the other hand, human chorionic gonadotrophin (hCG) production dropped as much as 50% while placental alkaline phosphatase (PLAP) elevated, regardless of the presence of cyanuric acid (Partanen, et al. 2012). Maternal oral dosing of melamine resulted in the detection of it in the foetuses (Chu, Chu and Chan, et al. 2010) (Chan, et al. 2011). Furthermore, it is proven that this placental transfer of melamine is in a dose-dependent fashion (Jingbin, et al. 2010).

2.2. Melamine Distribution

There are only a few studies reporting the distribution of melamine in the foetuses. A single oral bolus dose of melamine, around 5 times of the TDI, to the maternal rats in gestational day 20 revealed the xenobiotic can be measured in the foetal liver, brain, kidney, lung and heart, in the descending order of the area under the curve (AUC) of melamine in a toxicokinetic study, but the maximum concentration of melamine (C_{max}) was measured in the liver and heart instead (Chu, Chu and Ho, et al. 2013), which could be related to the higher blood flow to these two organs. Detection of melamine in the amniotic fluid was also reported (Chu, Chu and Chan, et al. 2010) (Chan, et al. 2011). Although the foetal kidney is not the organ for handling waste in the uterus, the excretion of melamine into the amniotic fluid enhances its continuous ingestion by the foetus, which may not be very well analysed by single bolus toxicokinetic studies. However, these toxicokinetic studies can reveal that melamine may not have the effect only on the kidney, but also on the other vital organs during foetal development (Figure 4).

2.3. Reproductive Toxicity

The incidence of the foetal death and litter size did not show a significant correlation with the dose (Kim, Lee and Lim, et al. 2011) (Stine, et al. 2014) (Chu, Tang, et al. 2017). In general, maternal melamine exposure did not show significant influence on the litter size and implantation loss in the studies employing either high dose or low dose of melamine (Kim, Lee and Lim, et al. 2011) (Chu, Tang, et al. 2017), with one exception of feeding the maternal rats with 1000 mg melamine/kg body weight/day (Stine, et al. 2014). The numbers of late resorption and still birth were noted as higher in the melamine groups on the study using low doses of melamine (Chu, Tang, et al. 2017). A dose-dependent decrease of the number of somites of melamine-receiving foetuses at gestational day 10.5 of the group of a 2-day daily maternal dose of 50 mg melamine/kg body weight (Chu, Tang, et al. 2017), a significant decrease in the crown-rump length of foetuses after 10 days of administration of 1000 mg melamine/kg body weight/day (Stine, et al. 2014) and a significantly lower birth weight of the neonates of the dams received 800 mg melamine/kg body weight/day (Kim, Lee and Lim, et al. 2011) suggested a possible growth retardation in uterus brought about by melamine. External anomalies of the neonates were rare. Size and weight of placentae of the melamine groups studied were comparable to the controls. Nevertheless, maternal mortality was possible: a high maternal death rate was noted in the group receiving 1600 mg/kg body weight/day of melamine (Kim, Lee and Lim, et al. 2011). Co-administration of melamine and cyanuric acid in 1:1 ratio showed more undesirable effects to the dams, e.g., lower birth weight of the neonates and higher death rate of the dams, even in much lower doses (Kim, Lee and Baek, et al. 2013). Also, the maternal weights of heart, kidney and adrenal gland relative to the body weight were higher in the melamine groups (Kim, Lee and Lim, et al. 2011) (Stine, et al. 2014) (Chu, Tang, et al. 2017).

2.4. Melamine Toxicity to the Foetal Kidneys

In a reproductive toxicology study published in 2017, size and weight of the neonatal rat kidneys significantly increased after maternal melamine exposure. The ill-famous melamine crystal was not observed in the postnatal day 1 rat kidneys after receiving melamine through the maternal exposure of maximum dose of 50 mg/kg body weight/day, around 10 times of TDI, for the whole pregnancy or the maturation developmental stage, i.e., gestational day 16.5 to birth. However, there was a general increase in the incidences of blood-congested glomeruli, interstitial blood congestion, dilated tubules and interstitial oedema in these neonatal kidneys. Chemically-induced tubulopathy is much more likely to result in simple hyperplasia than atypical hyperplasia. In these foetal rats exposed to maternal melamine, extensive simple hyperplasia with dilated tubules, without atypia, were occasionally observed. The increase in the percentage of the more matured glomeruli suggested that there may be an accelerated growth of the developing foetal kidneys under the influence of the maternal intake of melamine (Chu, Tang, et al. 2017).

One important point to note is that the neonatal rat is less developed than the newborn human. When making interpretation of these results, the findings of the neonatal rat kidneys should be correlated to that of the 2nd trimester of the human pregnancy while the findings of the postnatal 2-week rats may be better correlated to that of the 3rd trimester human (Tufro-McReddie, et al. 1995). No matter how, any injury to the developing kidneys may elicit impairment of the kidney function in the later life. It is well known that the low nephron endowment at birth is associated with hypertension (Brenner, Garcia and Anderson 1988) (Bagby 2007) and cardiovascular dysfunction (Moritz, et al. 2009) in adult life whereas the high nephron endowment can be protective against hypertension (Walker, et al. 2012). High nephron number of the offsprings of rats receiving high fat diet did not show observable nephropathy 9 months after birth (Hokke, et al. 2016), but long-term observation is lacking. If advanced maturity of the developing glomeruli will lead to high nephron endowment, the observation in the melamine-treated rat neonates may suggest a rescue mechanism been triggered in the developing foetus against the effect of melamine, at the point of observation.

2.5. Melamine Toxicity to the Foetal Skeleton

Development of the foetal skeleton was studied in detail under the effect of maternal exposure of melamine by a Korean group, by administration of around 160 times of the TDI of melamine to maternal rats for 15 days. The authors reported that there was a decrease in foetal body weight and an increased incidence of delayed foetal ossification. This study showed more observations of the skeletal variations including enlarged fontanel, incomplete ossification of supraoccipotalis, incomplete ossification of interparietal, bipartite ossification of sternebra and incomplete ossification of pubis. The number of ossification centres in sternebra, metacarpals and sacral and caudal vertebra were all decreased (Kim, Lee and Lim, et al. 2011). So as to avoid maternal death, a much lower dose has to be employed when melamine and cyanuric acid were administered in 1:1 ratio. Administering a 15-day daily dose of melamine, around 6 times of TDI, together with cyanuric acid, i.e., 30 + 30 mg/kg body weight, did not trigger any skeletal malformation or delay in skeletal growth of the rat foetuses as demonstrated in their previous study using only melamine as the challenge (Kim, Lee and Baek, et al. 2013).

Although there are documentations of melamine causing possible delay in the foetal growth, the foetal bone development and the nephropathology, melamine, with or without cyanuric acid, is not concluded as teratogenic.

2.6. Physiology Influencing Maternal Toxicology

Physiological changes during pregnancy may increase the risk of the contact of a xenobiotic; these changes include gastrointestinal emptying, cardiac output, tidal volume of lung in response to the need of the growing foetus. The *pKa* value of melamine is 5.0 (Jang, et al. 2009) which implies a less efficient absorption in the stomach is expected, due to the increased gastric emptying rate during pregnancy discourages the absorption of weak acids (Heikkilä and Erkkola 1994). Increase in blood volume also contributes to the distribution of the xenobiotic. For the 32nd week pregnancy in human, there is an increase of 45–60% in plasma volume and a decrease in about 20% of vascular resistance (Gei and Hankins 2001), which may favour the peripheral absorption of the xenobiotic. With this increase in blood volume, the concentration of plasma albumin decreases as a consequence (Krauer, Dayer and Anner 1984). Since melamine can bind to the hydrophilic residue on the surface of bovine serum albumin (Yan, et al. 2010), its distribution in pregnancy may be altered. In normal pregnancy, the increase in renal blood flow and glomerular flow rate (GFR) can readily elevate the risk of renal infection and hydronephrosis (Blackburn 2003); stress caused by melamine may further aggravate the maternal condition.

Moreover, the proper function of placenta is also crucial for the development of the foetus (Blackburn 2003). The decrease in hCG demonstrated in the melamine placental perfusion study (Partanen, et al. 2012) suggested a risk of spontaneous start of labour (Edelstam, et al. 2007). Furthermore, since the preterm neonates are more susceptible to nephrocalcinosis (Schell-Feith, Kist-van Holthe and van der Heijden 2010), the melamine challenge may add insult to injury to the newborn kidneys. A toxicokinetic study revealed the high apparent volume of distribution (Vz/F) in mid-gestation suggested the ingested melamine is more likely to be distributed to the periphery, either to the foetuses or to other maternal organs. The apparent clearance (Cl/F) that was higher in early gestation and dropped to a level comparable to the non-pregnant rats agrees with the changes in GFR along the course of pregnancy (Chu, Chu and Ho, et al. 2013). Regarding the serum toxicokinetics of melamine, the xenobiotic seemed to be cleared more efficiently in the earlier gestation; however, a higher incidence of crystals found in the maternal kidneys was reported in the same period of time of gestation (Chu, Tang, et al. 2017). Kidney stones are uncommon in normal pregnancy, but the incidence of hydronephrosis and dilated tubules is higher (Miller and Kakkis 1982) (Rodriguez and Klein

1988) (Cheung and Lafayette 2013). Maternal contact of melamine may then further complicate pregnancy.

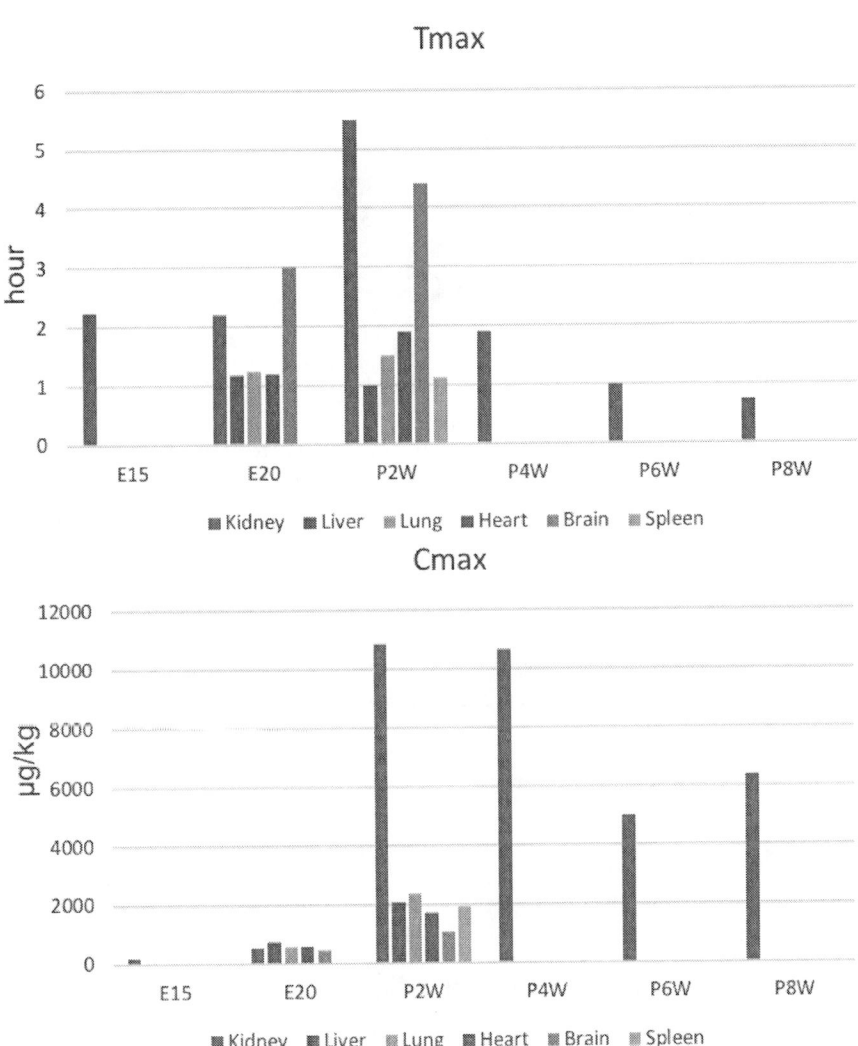

Figure 4. (Continued)

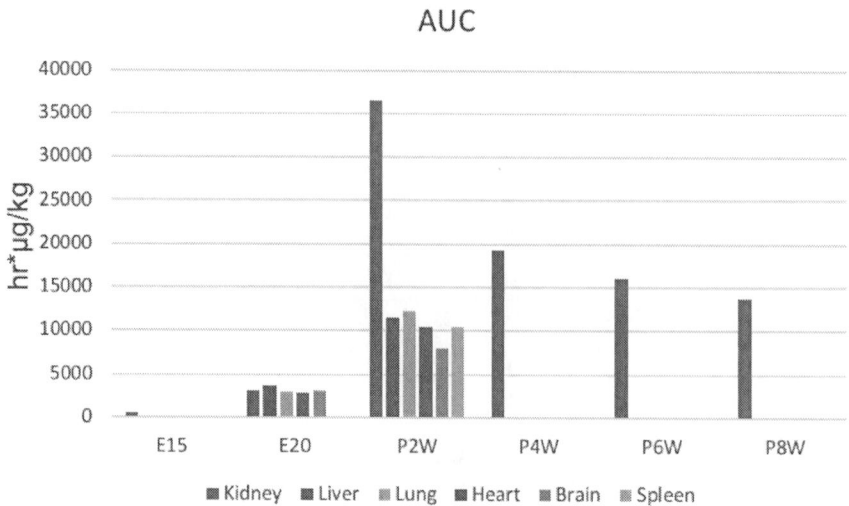

Figure 4.3. Toxicological data of foetal and postnatal exposure of melamine. T_{max} (time of maximum concentration), C_{max} (maximum concentration) and AUC (area under the curve) of melamine in foetal and postnatal organs. Data from Chu et al. 2013.

3. *Ex Utero* Contact of Melamine

3.1. Lactational Transfer of Melamine

Infants feeding the maternal milk is so natural in mammals. Milk provides nutrients for growth, antibodies for defense, and xenobiotics from the maternal diets for training of the infantile immune systems. Toxic substances may eventually be introduced to the immature newborns by means of ingestion.

As milk of the monogastric animals are not a common source of commercial milk for human consumption, studies of the lactational transfer of melamine were often carried out in ruminants. Rat is the only monogastric animal model for lactational transfer of melamine. The maximum concentration of melamine detected in rat's milk was detected in around the 3rd hour after the single bolus administration to near-term rats (Chu, Chu and Chan, et al. 2010) (Chan, et al. 2011). Estimation of the percentage transferred was not performed in these 2 experiments.

Feeding cows and goats melamine resulted in the detection of the xenobiotic in their milk (Cruywagen, et al. 2009) (Battaglia, et al. 2010) (Shen, et al. 2010) (Baynes, et al. 2010) (P. Sun, J. Wang, et al., Residues of melamine and cyanuric acid in milk and tissues of dairy cows fed different doses of melamine 2011) (P. Sun, J. Wang, et al., Pathway for the elimination of melamine in lactating dairy cows 2012). The half-life of melamine in goat's milk was reported to be around 9 hours, and it took 3 days to wash out the melamine in milk after a single oral dose of 40 mg/kg body weight (Baynes, et al. 2010). Melamine was detected in the milk as soon as 6 hours after feeding cows melamine-laced diets (Battaglia, et al. 2010) and the transfer efficiency was about 2–3% (Cruywagen, et al. 2009) (Battaglia, et al. 2010). Melamine content in the diet did not have influence on the transfer efficiency (Shen, et al. 2010) (Battaglia, et al. 2010).

These current researches revealing the maternal exposure of melamine can deduce the possibility of the presence of the xenobiotic in the human breast milk, which may serve as the sole food source for the newborns as breastfeeding is now re-introduced to the public as a healthier choice that compared to infant formulae. Although unlike human, cows and goats are ruminants, these studies are essential since milk of cow and goat serves mainly as the protein source for formulae for infants, children, pregnant and lactating women, also as ingredients of snacks. The relevant information provided can help the regulatory organisations to decide the upper limit of melamine in the animal feeds, or whether or not it is appropriate to be a filler at all.

3.2. Melamine-Contaminating Infant Formulae

From the evidence provided by the above-mentioned studies, we know that deploying melamine as a filler for animal feeds can bring melamine to the milk produced. The adulteration of infant formulae by melamine during the scandal was not as complicated, but simply an economically-driven direct lacing by the contaminant.

Besides the Chinese AQSIQ report on the melamine content in the infant formulae in the Chinese market and the report of the highest melamine content ever detected by Jia et al. (2009), a few more groups of researchers have measured the sampled formula obtained during the time of the scandal (Jia, et al. 2009). A study in Hangzhou, China showed the highest level of melamine measured in their study was 2563 mg/kg (Zhang, et al. 2009) while another study in Gansu Province, China showed the mean of melamine content of the brand of infant formula that had the highest adulteration was 1673.58 mg/kg (Wu, et al. 2009). Although since the cessation of the melamine crisis reports on the news by the end of the last decade, there has been scarce follow-up about its existence on the shelf, Hassani et al. (2013) has revealed that there is still an adulteration milk by melamine, in the range of 1.5–30.32 µg/g of milk powder and 0.11–1.48 µg/ml of liquid milk in the Iranian market (Hassani, et al. 2013), while García Londoño et al. (2018) reported a 9 in 40 milk powder samples collected in Uruguay contained melamine in the range of 0.017–0.082 mg/kg (García Londoño, et al. 2018) and other food products in the US (Zhu and Kannan 2019). Whether or not these limited number of reports is due to the lack of surveillance or the true absence of melamine in food, it still proves that the risk is not over.

3.3. Melamine Distribution

The toxicokinetic study carried out by Chu et al. (2013) compared the distribution of melamine between the near-term foetuses and the neonates of rats contacting melamine *in utero* and *ex utero* respectively. The postnatal 2-week rats, whose kidneys start to eliminate waste, were the most vulnerable to the melamine toxicity, as shown by their highest AUC in serum and kidney when compared to the more developed postnatal rats at the age of 4–8 weeks. Furthermore, as mentioned in the previous session, the distribution of melamine to the kidney of the near-term rat receiving melamine through the placenta did not show much different among the organs studied; rather more melamine was distributed to the foetal liver and brain. In contrast, in the postnatal 2-week rats receiving melamine through ingestion showed the distribution of melamine to the kidney was at least 3 times more than that of the other organs studied. Again, when the neonatal kidneys adopt the responsibility of ultrafiltration, it makes possible the melamine concentration and accumulation there, leading to the possibility of stone formation and in turn the injury to the nephrons (Figure 4) (Chu, Chu and Ho, et al. 2013).

3.4. Melamine Toxicity to the Neonatal Kidneys

The earliest paper published about the clinical cases of the melamine-affected infants in the scandal came in 2008. From the recruited 2186 children under the age of 3: 1329 males and 857 females, the authors have estimated the total amount of the most seriously-adulterated Sanlu infant formulae consumed by the infants and correlated the amount consumed with the prevalence of the urinary tract stones. The prevalence rose from 11.4% to 37.5% between the amount consumed from >400 g to 25600–76000 g. Nephrolithiasis (16.6%) and urolithiasis (16.5%) were the most frequently identified; their prevalence in male and female was comparable. On the contrary, hydronephrosis, ureteral stones and bladder stones seemed to be sex-related, upsetting more males then females (Shi, et al. 2012). Another study has shown that the incidence of melamine-associated kidney stone was also sex-biased: male had about 2-fold risk higher that female (Lu, et al. 2011). In general occasions, a higher incidence of urinary tract problems is often found in male. The size of these stones showed a correlation with the amount of melamine received instead of the duration (Hu, Lu, et al. 2010) and the presence of calcium urolithiasis was found to be related to the urinary concentration of melamine (Liu, et al. 2011). Nonetheless, another study in Taiwan correlated the risk to the duration of consumption and the estimated melamine levels instead (Wang, et al. 2009). The reason for this discrepancy may be due to the fact that the contamination of food by melamine in Hong Kong and Taiwan was not as serious as in China (Centre for Food Safety, HK 2018) (Lam, et al. 2008) (Wang, et al. 2009). Importantly, Li et al. (2010) have reminded us that children ingesting melamine at a level lower than the WHO recommendation, i.e., 0.2 mg/kg, still had 1.7 times higher risk of melamine-induced nephrolithiasis (Li, Jiao, et al. The risk of melamine-induced nephrolithiasis in young children starts at a lower intake level than recommended by the WHO 2010).

3.5. Melamine Toxicity to the Neonatal Brain

Toxicokinetic studies showed the distribution of melamine in brain was higher in the prenatal rats than in the kidney and it was also detected in the brain of the postnatal brain (Chu, Chu and Ho, et al. 2013). A paper published in 2011 has shown the *ex utero* effect of melamine onto the learning ability of postnatal 3-week-old rats (Yang, et al. 2011). Both the Morris water maze test and the long-time potentiation results showed the impairment of the spatial learning and memory abilities, which reflect the function of hippocampus (Clark, Broadbent and Squire 2007). Excitability of rat hippocampal CA1 pyramidal neurons increased in the melamine-treated brain slice culture (Yang, et al. 2010). An acute low-dose of melamine impaired long-term potentiation without having effects on the long-term depression in another study using rat hippocampus. Basal synaptic transmission in the Schaffer collateral-CA1 pathway decreased in hippocampus slice culture and the effect was larger in the postnatal 1-week-old than 2–3-month-old rats (Yang, et al. 2012). Study of the distribution of melamine in the brain was only carried out in adult rat brain: hippocampus has the 2nd highest maximum concentration of melamine detected, after brain stem (Wu, et al. 2009).

Later, An and Zhang (2016) has performed another experiment comparing the long-term effect of melamine exposure pre- and post-natally for 20 days. No matter the spatial cognition and synaptic impairments, the paired-pulse facilitation ratio and post-tetanic potentiation or the long-term potentiation, the detrimental effect of melamine was larger in the prenatally-affected group, even though the melamine level in postnatal serum or hippocampus of the prenatally-exposed rats had returned to a level comparable to that of the control group. The weight of hippocampus relative to the brain of the prenatally-exposed group was significantly lower than the control group. These pieces of evidence suggest that the neurotoxic effect of melamine can be long-lasting and the time of exposure can determine the impact (An and Zhang 2016).

3.6. Melamine Toxicity to the Neonatal Liver

There is not yet a detail report of the hepatotoxicity of melamine about *in utero* or *ex utero* contact of the xenobiotic, only clinical descriptions of the affected human infants were available, ranging from hepatomegaly, elevated aspartate aminotransferasemia, gallstone, liver lesions (Zhang, et al. 2009), gradual progressive jaundice, abdominal distention, hepatic intumesce and bilirubin abnormality (Hu, Wang and Zhang, et al. 2012). Biochemical study showed that 3 of the haem peroxidases, namely horseradish peroxidase, lactoperoxidase and cyclooxygenase-1 were inhibited by melamine, which may impair haematopoiesis, host defense mechanism and the production of prostaglandins (Vanachayangkul and Tolleson 2012). In accord with the above-mentioned study, haematopoietic cell lineage was found to be among the top 20 enriched KEGG pathways identified by the iTRAQ technique using adult mice liver cells. In the same study, the authors suggested the proteins involved in immune and inflammatory function, unfolded proteins response in endoplasmic reticulum, DNA damage, and the apoptosis of liver cells may also be the targets of melamine hepatotoxicity (Yin, et al. 2019).

4. FOLLOW UP

Clinical findings of the follow-up of the infants suffered in the melamine scandal are also available. Most of them focused on the outcome of the interventions made but little is about the other assessments of the children. Two 5-year follow-up studies in Gansu Province, China showed that the incidence of AKI was related to age, the duration of the intake, and the concentration of melamine presence in the infant formulae, assessed by the brands they took. Blood urea nitrogen (BUN) and creatinine (Cr) of all the AKI patients returned to normal in the 18-month follow-up and renal calculi were found in none of the patients in the 5-year follow-up, even though some of them were still detected in the 3-year follow-up (He, et al. 2014). Zou et al. (2013) revealed that a period of 2 years may not be long enough for the disappearance of the larger urinary stones; moreover, hydronephrosis and gallstones may persist even the stones were passed or removed, and BUN and Cr had returned to normal (Zou, et al. 2013). Chronic urinary stones increased the rate of glomerular injury (Gao, et al. 2016). Both the patient groups receiving conservative or surgical treatments had the urolithiasis resolved in 5 years (Chang, Wu, et al. 2017). Height and weight of the melamine-affected children were reported to be less than the normal average (Zou, et al. 2013) (He, et al. 2014), but He et al. suspected that the stopping of taking milk after the scandal due to the lack of confidence may be one of the reasons for the slowed-down growth (He, et al. 2014).

5. Low-Dose Contamination of Melamine

Excretion of the undigested melamine, leakage from the melamine-containing materials and degradation of the its precursors can bring melamine to where we have not expected it to be. These contaminations are usually unintentional and difficult to regulate. A very recent study published this year in 2019 reminds us that melamine and its derivatives, namely ammeline, ammelide and cyanuric acid (Jutzi, Cook and Hütter 1982), which was also found in melamine-contaminated food but in lower amounts, do still exist in the foodstuffs in the United States nowadays. The report showed that melamine and its derivatives were detected in meat, fish and seafood, cereals, vegetables, cooking oil, beverages, dairy products, as well as food packaging and animal feed, despite the small amount (Zhu and Kannan 2019), i.e., almost everywhere. The highest average sum of melamine and its 3 derivatives detected was 23.6 ng/g, which is well below the safety level of melamine published by the US FDA is 2.5 mg/kg in food (FDA 2008). Maleki et al. (2018) has also told us that 45 out of the 69 sampled infant/follow-up formulae on the Iranian market were still contaminated by melamine, in which 10 samples had the melamine level exceeding the recommended safety level set by the WHO for infant food (Maleki, et al. 2018). An earlier study in 2015 reported that nearly half of the protein supplements in the South African market contained melamine and the median estimated concentration was 6.031 mg/kg, which is higher than the safety recommendation (Gabriels, et al. 2015). In consideration of the intake of the supplement, the authors commented that it is still under the human tolerable daily intake (TDI) suggested by the World Health Organisation (WHO) in 2008, i.e., 0.2 mg/kg body weight/day (FDA 2008) (WHO 2012).

Some of the melamine presence in foodstuff can be the consequence of the farm animals ingesting the melamine-laced feed. Besides milk, tainted feed can give rise to tainted eggs (Lü, et al. 2009) (Dong, et al. 2010) (Gao, et al. 2010) (Zhang, Guo and Wang 2012) (Suchý, et al. 2014), tainted muscles of chicken, pig and fish (Lü, et al. 2009) (Dong, et al. 2010) (Gallo, et al. 2012) (Phromkunthong, et al. 2013) (Suchý, et al. 2014) (Wang, et al. 2014), also tainted internal organs of chicken and cow (Lü, et al. 2009) (Dong, et al. 2010) (P. Sun, J. Wang, et al., Residues of melamine and cyanuric acid in milk and tissues of dairy cows fed different doses of melamine 2011) (Suchý, et al. 2014). Withdrawal time depends largely on the level of melamine laced in the animal feeds. Studies showed that 1 week should be allowed for the withdrawal of melamine for birds for the presence of up to 1000 mg melamine/kg feed (Lü, et al. 2009) (Dong, et al. 2010) (Zhang, Guo and Wang 2012), and a 5-day period for an addition of 500 mg melamine/kg for pigs (Wang, et al. 2014). However, a single dose of melamine took 9.5 hours to reach the highest concentration in goat's milk and may take 60 hours to be cleared to an acceptable level (Baynes, et al. 2010). Considering the safety guidelines recommended by WHO that the limit of presence of melamine in infant formula at 1 mg/kg level, Shen et al. (2010) suggested that the milk yielded from a cow with daily intake of 312.7 mg of melamine should not be used as the source for the producing infant formulae (Shen, et al. 2010) while Battaglia et al. (2010) has also found that daily intake of 500 mg of melamine produces milk with a melamine level similar to the WHO guideline (Battaglia, et al. 2010). The clearance rate of melamine in cow's milk was about 17 g per day (Cruywagen, et al. 2009) (Battaglia, et al. 2010). Sun et al. (2012) suggested that in fact more than half of the melamine ingested by a cow was eliminated in urine and feces, a little was converted into cyanuric acid and a large amount into other metabolite, which no qualification was performed (P. Sun, J. Wang, et al., Pathway for the elimination of melamine in lactating dairy cows 2012). No cyanuric acid was detected in the resultant milk samples (P. Sun, J. Wang, et al., Residues of melamine and cyanuric acid in milk and tissues of dairy cows fed different doses of melamine 2011) (P. Sun, J. Wang, et al., Pathway for the elimination of melamine in lactating dairy cows 2012); therefore, the risk of the more detrimental melamine-cyanuric acid crystal can be reduced if there is no co-adulteration of the animal feed by cyanuric acid.

Melamine ware or melaware is a common kitchenware for its durability, especially popular for children's tableware. Heating acidic food in these melaware and melamine-formaldehyde resin coated cans and jar lids can lead to leakage into food (Sugita, Ishiwata and Yoshihira 1990) (Bradley, Boughtflower, et al. 2005) (Lund and Petersen 2006) (Bradley, Castle, et al. 2011) (Chik, et al. 2011), but all fell below the specific migration limit (SML) of 30 mg/kg or 5 mg/dm^2 at the time of the assessment (EC 2002) (EC 2003). Later in 2011, European Union has lowered the SML of melamine to 2.5 mg/kg (EU 2011); therefore, some of them failed to comply with the regulation. Mannoni et al. (2017) have shown that, with the lowered SML, not all the new melaware sampled from the Italian market has the melamine migration below the safety level; even more, repeated use of the old melaware may result in higher migration (Mannoni, et al. 2017). The authors suggested that diffusion may not be the reason for the observation but the possibility of hydrolytic degradation of the polymer itself. A reputable German consumer organisation and foundation Stiftung Warentest has reminded the consumers that these resuable coffee cups although made of bamboo fibers, are in fact glued by melamine resin. Their recent test showed only 1 of the 12 samples did not show migration of melamine or formaldehyde to the testing drink (DW 2019). However little it is, these studies have provided evidence that the public is still ingesting a certain amount of melamine and its derivatives every day.

Low-dose melamine contamination seems to be unavoidable. Should it be possible to add the screening for melamine as a routine? Although WHO has set the safety level as 0.2 mg/kg per day, Li et al. (2010) has shown the 1.7 times higher chance of having nephrolithiasis if exceeded (Li, Jiao, et al. The risk of melamine-induced nephrolithiasis in young children starts at a lower intake level than recommended by the WHO 2010). Using the benchmark dose analysis, Hsieh et al. (2009) have suggested the TDI be further lowered to 0.0081 mg/kg body weight/day, instead of the 0.063 mg/kg body weight/day recommended by the FDA (Hsieh, et al. 2009).

CONCLUSION

A comprehensive assessment on the safety of a chemical can take a long time; assessing the safety of a chemical on a multi-generation basis needs even longer. The Organisation for Economic Co-operation and Development (OECD) has publish a list of guidelines for the testing chemicals. Section 4 addresses to the health effects that covers the prenatal development toxicity study (Test No 414), one and two-generation reproduction toxicity studies (Test Nos 415/416), reproduction/development toxicity screening test (Test No 421), combined repeated dose toxicity study with the reproduction-development toxicity screening test (Test No 422), developmental neurotoxicity study (Test No 426), extended one-generation reproductive toxicity study (Test No 443) (OECD n. d.). These guidelines help researchers to develop their chemical evaluation studies, incorporated with their own research interests.

To reduce the possible exposure to melamine, which is now classified as a Group 2B carcinogen, i.e., 'possibly carcinogenic to humans,' by the International Agency for Research on Cancer, WHO (IARC 2019), or other chemical contamination in food, a thorough risk management has to be carried out. First of all, the assessment of the toxicity of the chemical, and the assessment of the contact by human, especially by the vulnerable groups, including the source of the contamination, the food intake amount and frequency have to be studied in great detail. Then the policy-making organisations have to design a plan on how to control the risk, e.g., to prevent the toxic chemicals from going into the food chain, to set up guidelines of the upper limit for the human contact, and to enforce the implementation of the policy by every stakeholders. Prompt recall once a health risk is identified is critical. Moreover, communication between the scientific evidence providers and the policy makers is very important, but the communication with the consumers is the ultimate goal. With the unavoidable global trade and the growing complexity of the food supply chain, transparency and traceability of the food supply and the integrity of the risk managing team is essential to curb the unnecessary risks carried in food.

Just like we have been using plastics for decades, we start to realise that they now exist in new forms, the micro- or nano-plastics, even in the virgin lands and in our own feces. Have we ever thought of how melamine disappear from our household, and where the melamine foam goes after cleaning the kitchen? Will it come back to us? And how would it be like? In 2016, The Frank R. Lautenberg Chemical Safety for the 21st Century Act is passed in the United States aiming to evaluate the safety of all chemicals on the market, the existing ones and the new, under the conditions of use whether they present any unreasonable risks. The evaluation bases solely on the existing scientific evidence, without balancing with the non-risk factors, e.g., cost and benefits. Special consideration for the effects onto the more vulnerable groups, e.g., the infants, the children and the pregnant women should be taken into account (USEPA n. d.). It will definitely take a very long road to evaluate thousands of chemicals existing in the market, if not millions, but it is good start with a kind heart.

REFERENCES

An, L, and T Zhang. 2016. "Comparison impairments of spatial cognition and hippocampal synaptic plasticity between prenatal and postnatal melamine exposure in male adult rats." *Neurotox Res* 29(2):218-29.

Bagby, SP. 2007. "Maternal nutrition, low nephron number, and hypertension in later life: pathways of nutritional programming." *J Nutr* 137(4):1066-72.

Battaglia, M, CW Cruywagen, T Bertuzzi, A Gallo, M Moschini, G Piva, and F Masoero. 2010. "Transfer of melamine from feed to milk and from milk to cheese and whey in lactating dairy cows fed single oral doses." *J Dairy Sci* 93(11):5338-47.

Baynes, RE, B Barlow, SE Mason, and JE Riviere. 2010. "Disposition of melamine residues in blood and milk from dairy goats exposed to an oral bolus of melamine." *Food Chem Toxicol* 48(8-9):2542-6.

Blackburn, S. 2003. "Renal system and fluid and electrolyte homeostasis." In *S. Blackburn, Maternal, fetal, & neonatal physiology: a clinical perspective, 2nd ed*, by S Blackburn, 370-411. St Louis, MO: Saunders.

Bradley, EL, L Castle, JS Day, and J Leak. 2011. "Migration of melamine from can coatings cross-linked with melamine-based resins, into food simulants and foods." *Food Addit Contam Part A Chem Anal Control Expo Risk Assess* 28(2):243-50.

Bradley, EL, V Boughtflower, TL Smith, DR Speck, and L Castle. 2005. "Survey of the migration of melamine and formaldehyde from melamine food contact articles available on the UK market." *Food Addit Contam* 22(6):597-606.

Brenner, BM, DL Garcia, and S Anderson. 1988. "Glomeruli and blood pressure. Less of one, more the other?" *Am J Hypertens* 1:335–47.

Brown, CA, KS Jeong, RH Poppenga, B Puschner, DM Miller, AE Ellis, KI Kang, S Sum, AM Cistola, and SA Brown. 2007. "Outbreaks of renal failure associated with melamine and cyanuric acid in dogs and cats in 2004 and 2007." *J Vet Diagn Invest* 19(5):525-31.

Centre for Food Safety, HK. 2018. *Unsatisfactory results of testing of melamine in food samples (As at 2 Dec 2008)*. https://www.cfs.gov.

hk/english/whatsnew/whatsnew_fstr/files/melamine_dec/List_of_unsatisfactory_food_samples.pdf.

Chan, JY, CM Lau, TL Ting, TC Mak, MH Chan, CW Lam, CS Ho, CC Wang, TF Fok, and KP Fung. 2011. "Gestational and lactational transfer of melamine following gavage administration of a single dose to rats." *Food Chem Toxicol* 49(7):1544-8.

Chang, H, G Wu, Z Yue, J Ma, and Z Qin. 2017. "Melamine Poisoning Pediatric Urolithiasis Treatment in Gansu, China 5-Year Follow-up Analysis." *Urology* 109:153-158.

Chang, H, X Shi, W Shen, W Wang, and Z Yue. 2012. "Characterization of melamine-associated urinary stones in children with consumption of melamine-contaminated infant formula." *Clin Chim Acta* 413(11-12):985-91.

Chen, KC, CW Liao, FP Cheng, CC Chou, SC Chang, JH Wu, JM Zen, YT Chen, and JW Liao. 2009. "Evaluation of subchronic toxicity of pet food contaminated with melamine and cyanuric acid in rats." *Toxicol Pathol* 37(7):959-68.

Cheng, G, N Shapir, MJ Sadowsky, and LP Wackett. 2005. "Allophanate hydrolase, not urease, functions in bacterial cyanuric acid metabolism." *Appl Environ Microbiol* 71(8):4437-45.

Cheung, KL, and RA Lafayette. 2013. "Renal physiology of pregnancy." *Adv Chronic Kidney Dis* 20(3):209-14.

Chik, Z, DE Haron, ED Ahmad, H Taha, and AM Mustafa. 2011. "Analysis of melamine migration from melamine food contact articles." *Food Addit Contam Part A Chem Anal Control Expo Risk Assess* 28(7):967-73.

China Daily. 2008. *Melamine in more milk.* Sep 17. http://www.chinadaily.com.cn/china/2008-09/17/content_7032353.htm.

Chu, CY, KO Chu, CS Ho, SS Kwok, HM Chan, KP Fung, and CC Wang. 2013. "Melamine in prenatal and postnatal organs in rats." *Reprod Toxicol* 35:40-7.

Chu, CY, KO Chu, JY Chan, XZ Liu, CS Ho, CK Wong, CM Lau, et al. 2010. "Distribution of melamine in rat foetuses and neonates." *Toxicol Lett* 199(3):398-402.

Chu, CY, LY Tang, L Shum, AS Li, KP Fung, and CC Wang. 2017. "Adverse reproductive effects of maternal low-dose melamine exposure during pregnancy in rats." *Environ Toxicol* 32(1):131-138.

Cianciolo, RE, K Bischoff, JG Ebel, TJ Van Winkle, RE Goldstein, and LM Serfilippi. 2008. "Clinicopathologic, histologic, and toxicologic findings in 70 cats inadvertently exposed to pet food contaminated with melamine and cyanuric acid." *J Am Vet Med Assoc* 233(5):729-37.

Clark, RE, NJ Broadbent, and LR Squire. 2007. "The hippocampus and spatial memory: findings with a novel modification of the water maze." *J Neurosci* 27(25):6647-54.

Codex Alimentarius Commission. 2016. *Standard for infant formula and formulas for special medical purposes intended for infants CODEX STAN 72 – 1981.* http://www.fao.org/fao-who-codexalimentarius/sh-proxy/en/?lnk=1&url=https%253A%252F%252Fworkspace.fao.org%252Fsites%252Fcodex%252FStandards%252FCXS%2B72-1981%252FCXS_072e.pdf.

Conlan, S, HH Kong, and JA Segre. 2012. "Species-level analysis of DNA sequence data from the NIH Human Microbiome Project." *PLoS One* 7(10):e47075.

Cruywagen, CW, MA Stander, M Adonis, and T Calitz. 2009. "Hot topic: pathway confirmed for the transmission of melamine from feed to cow's milk." *J Dairy Sci* 92(5):2046-50.

Dai, ZL, XL Li, PB Xi, J Zhang, G Wu, and WY Zhu. 2012. "Metabolism of select amino acids in bacteria from the pig small intestine." *Amino Acids* 42(5):1597-608.

Dodge, AG, LP Wackett, and MJ Sadowsky. 2012. "Plasmid localization and organization of melamine degradation genes in Rhodococcus sp. strain Mel." *Appl Environ Microbiol* 78(5):1397-403.

Dong, XF, SY Liu, JM Tong, and Q Zhang. 2010. "Carry-over of melamine from feed to eggs and body tissues of laying hens." *Food Addit Contam Part A Chem Anal Control Expo Risk Assess* 27(10):1372-9.

DW. 2019. *'Keep your hands off' bamboo coffee cups, German consumer group warns.* Jul 23. https://www.dw.com/en/keep-your-hands-off-bamboo-coffee-cups-german-consumer-group-warns/a-49713624.

Eaton, RW, and JS Karns. 1991. "Cloning and analysis of s-triazine catabolic genes from Pseudomonas sp. strain NRRLB-12227." *J Bacteriol* 173(3):1215-22.

EC. 2002. "Commission Directive 2002/72/EC relating to plastics materials and articles intended to come into contact with foodstuffs." *Official Journal of the European Communities* L220:18–58.

EC. 2003. "Corrigendum to Commission Directive 2002/72/EC of 6 August 2002 relating to plastic materials and articles intended to come into contact with foodstuffs." *Official Journal of the European Union* L 39:1–42.

Edelstam, G, C Karlsson, M Westgren, C Löwbeer, and ML Swahn. 2007. "Human chorionic gonadatropin (hCG) during third trimester pregnancy." *Scand J Clin Lab Invest* 67(5):519-25.

El-Sayed, W, A El-Baz, and A Othman. 2006. "Biodegradation of melamine formaldehyde by Micrococcus sp. strain MF-1 isolated from aminoplastic wastewater effluent." *Int Biodeterior Biodegrad* 57(2):75–81.

EU. 2011. *Commission Regulation (EU) No 1282/2011 of 28 November 2011, amending and correcting Commission Regulation (EU) No 10/2011 on plastic materials and articles intended to come into contact with food.* Dec 10. https://eur-lex.europa.eu/legal-content/EN/TXT/PDF/?uri=CELEX:32011R1282&from=EN.

FDA. 2008. *Interim safety and risk assessment of melamine and its analogues in food for humans.* https://www.federalregister.gov/documents/2008/11/13/E8-26869/interim-safety-and-risk-assessment-of-melamine-and-its-analogues-in-food-for-humans-availability.

Gabriels, G, M Lambert, P Smith, L Wiesner, and D Hiss. 2015. "Melamine contamination in nutritional supplements--Is it an alarm bell for the general consumer, athletes, and 'Weekend Warriors'?" *Nutr J* 14:69.

Gallo, A, T Bertuzzi, M Battaglia, F Masoero, G Piva, and M Moschini. 2012. "Melamine in eggs, plasma and tissues of hens fed contaminated diets." *Animal* 6(7):1163-9.

Gao, CQ, SG Wu, HY Yue, F Ji, HJ Zhang, QS Liu, ZY Fan, FZ Liu, and GH Qi. 2010. "Toxicity of dietary melamine to laying ducks:

biochemical and histopathological changes and residue in eggs." *J Agric Food Chem* 58(8):5199-205.

Gao, J, F Wang, X Kuang, R Chen, J Rao, B Wang, W Li, et al. 2016. "Assessment of chronic renal injury from melamine-associated pediatric urolithiasis: an eighteen-month prospective cohort study." *Ann Saudi Med* 36(4):252-7.

García Londoño, VA, M Puñales, M Reynoso, and S Resnik. 2018. "Melamine contamination in milk powder in Uruguay." *Food Addit Contam Part B Surveill* 11(1):15-19.

Gei, AF, and GD Hankins. 2001. "Cardiac disease and pregnancy." *Obstet Gynecol Clin North Am* 28(3):465-512.

González, J, B Puschner, V Pérez, MC Ferreras, L Delgado, M Muñoz, C Pérez, et al. 2009. "Nephrotoxicosis in Iberian piglets subsequent to exposure to melamine and derivatives in Spain between 2003 and 2006." *J Vet Diagn Invest* 21(4):558-63.

Hassani, S, F Tavakoli, M Amini, F Kobarfard, A Nili-Ahmadabadi, and O Sabzevari. 2013. "Occurrence of melamine contamination in powder and liquid milk in market of Iran." *Food Addit Contam Part A Chem Anal Control Expo Risk Assess* 30(3):413-20.

He, Q, Z Yue, X Tang, H Chang, W Wang, W Shi, Z Wang, and P Shang. 2014. "Risk factors for acute kidney injury (AKI) in infants with melamine-associated urolithiasis and follow-up: a multi-center retrospective analysis." *Ren Fail* 36(9):1366-70.

Heikkilä, A, and R Erkkola. 1994. "Review of beta-lactam antibiotics in pregnancy. The need for adjustment of dosage schedules." *Clin Pharmacokinet* 27(1):49-62.

Hokke, S, VG Puelles, JA Armitage, K Fong, JF Bertram, and LA Cullen-McEwen. 2016. "Maternal Fat Feeding Augments Offspring Nephron Endowment in Mice." *PLoS One* 11(8):e0161578.

Hsieh, DP, CF Chiang, PH Chiang, and CP Wen. 2009. "Toxicological analysis points to a lower tolerable daily intake of melamine in food." *Regul Toxicol Pharmacol* 55(1):13-6.

Hu, P, J Wang, B Hu, L Lu, and M Zhang. 2013. "Clinical observation of childhood urinary stones induced by melamine-tainted infant formula in Anhui province, China." *Arch Med Sci* 9(1):98-104.

Hu, P, J Wang, M Zhang, B Hu, L Lu, CR Zhang, and PF Du. 2012. "Liver involvement in melamine-associated nephrolithiasis." *Arch Iran Med* 15(4):247-8.

Hu, P, L Lu, B Hu, and CR Zhang. 2010. "The size of melamine-induced stones is dependent on the melamine content of the formula fed, but not on duration of exposure." *Pediatr Nephrol* 25(3):565-6.

IARC. 2019. *List of Classifications: Agents classified by the IARC Monographs, Volumes 1–124.* https://monographs.iarc.fr/list-of-classifications.

Jang, YH, S Hwang, SB Chang, J Ku, and DS Chung. 2009. "Acid dissociation constants of melamine derivatives from density functional theory calculations." *J Phys Chem A* 113(46):13036-40.

Jeong, WI, SH Do, DH Jeong, JY Chung, HJ Yang, DW Yuan, IH Hong, JK Park, MJ Goo, and KS Jeong. 2006. "Canine renal failure syndrome in three dogs." *J Vet Sci* 7(3):299-301.

Jia, XD, N Li, ZT Wang, YF Zhao, YN Wu, and WX Yan. 2009. "Assessment on dietary melamine exposure from tainted infant formula." *Biomed Environ Sci* 22(2):100-3.

Jingbin, W, M Ndong, H Kai, K Matsuno, and F Kayama. 2010. "Placental transfer of melamine and its effects on rat dams and fetuses." *Food Chem Toxicol* 48(7):1791-5.

Jutzi, K, AM Cook, and R Hütter. 1982. "The degradative pathway of the s-triazine melamine. The steps to ring cleavage." *Biochem J* 208(3):679-84.

Kim, SH, IC Lee, HS Baek, KW No, DH Shin, C Moon, SH Kim, SC Park, and JC Kim. 2013. "Effects of melamine and cyanuric acid on embryo-fetal development in rats." *Birth Defects Res B Dev Reprod Toxicol* 98(5):391-9.

Kim, SH, IC Lee, JH Lim, IS Shin, C Moon, SH Kim, SC Park, HC Kim, and JC Kim. 2011. "Effects of melamine on pregnant dams and embryo-fetal development in rats." *J Appl Toxicol* 31(6):506-14.

Krauer, B, P Dayer, and R Anner. 1984. "Changes in serum albumin and alpha 1-acid glycoprotein concentrations during pregnancy: an analysis of fetal-maternal pairs." *Br J Obstet Gynaecol* 91(9):875-81.

Lam, HS, PC Ng, WC Chu, W Wong, DF Chan, SS Ho, KT Wong, AT Ahuja, and CK Li. 2008. "Renal screening in children after exposure to low dose melamine in Hong Kong: cross sectional study." *BMJ* 337:a2991.

Li, G, S Jiao, X Yin, Y Deng, X Pang, and Y Wang. 2010. "The risk of melamine-induced nephrolithiasis in young children starts at a lower intake level than recommended by the WHO." *Pediatr Nephrol* 25(1):135-41.

Lim, LO, SJ Scherer, KD Shuler, and JP Toth. 1990. "Disposition of Cyromazine in Plants under Environmental Conditions." *J Agric Food Chem* 38: 860-4.

Liu, CC, CF Wu, BH Chen, SP Huang, W Goggins, HH Lee, YH Chou, et al. 2011. "Low exposure to melamine increases the risk of urolithiasis in adults." *Kidney Int* 80(7):746-52.

Lü, MB, L Yan, JY Guo, Y Li, GP Li, and V Ravindran. 2009. "Melamine residues in tissues of broilers fed diets containing graded levels of melamine." *Poult Sci* 88(10):2167-70.

Lu, X, J Wang, X Cao, M Li, C Xiao, T Yasui, and B Gao. 2011. "Gender and urinary pH affect melamine-associated kidney stone formation risk." *Urol Ann* 3(2):71-4.

Lund, KH, and JH Petersen. 2006. "Migration of formaldehyde and melamine monomers from kitchen- and tableware made of melamine plastic." *Food Addit Contam* 23(9):948-55.

Ma, M, and D Bong. 2011. "Determinants of cyanuric acid and melamine assembly in water." *Langmuir* 27(14):8841-53.

Maleki, J, F Nazari, J Yousefi, R Khosrokhavar, and MJ Hosseini. 2018. "Determinations of Melamine Residue in Infant Formula Brands Available in Iran Market Using by HPLC Method." *Iran J Pharm Res* 17(2):563-570.

Mannoni, V, G Padula, O Panico, A Maggio, C Arena, and MR Milana. 2017. "Migration of formaldehyde and melamine from melaware and other amino resin tableware in real life service." *Food Addit Contam Part A Chem Anal Control Expo Risk Assess* 34(1):113-125.

Mast, RW, AR Jeffcoat, BM Sadler, RC Kraska, and MA Friedman. 1983. "Metabolism, disposition and excretion of [14C]melamine in male Fischer 344 rats." *Food Chem Toxicol* 21(6):807-10.

Melnick, RL, GA Boorman, JK Haseman, RJ Montali, and J Huff. 1984. "Urolithiasis and bladder carcinogenicity of melamine in rodents." *Toxicol Appl Pharmacol* 72(2):292-303.

Miller, RD, and J Kakkis. 1982. "Prognosis, management and outcome of obstructive renal disease in pregnancy." *J Reprod Med* 27(4):199-201.

Moritz, KM, RR Singh, ME Probyn, and KM Denton. 2009. "Developmental programming of a reduced nephron endowment: more than just a baby's birth weight." *Am J Physiol Renal Physiol* 296(1):F1-9.

New York Times. 2007. *Filler in Animal Feed Is Open Secret in China.* April 30. https://www.nytimes.com/2007/04/30/business/worldbusiness/30food.html.

Newton, GL, and PR Utley. 1978. "Melamine as a dietary nitrogen source for ruminants." *J Anim Sci* 47:1338–1344.

Nilubol, D, T Pattanaseth, K Boonsri, N Pirarat, and N Leepipatpiboon. 2009. "Melamine- and cyanuric acid-associated renal failure in pigs in Thailand." *Vet Pathol* 46(6):1156-9.

OECD. n.d. *OECD Guidelines for the Testing of Chemicals, Section 4.* https://www.oecd-ilibrary.org/environment/oecd-guidelines-for-the-testing-of-chemicals-section-4-health-effects_20745788.

Ogasawara, H, K Imaida, H Ishiwata, K Toyoda, T Kawanishi, C Uneyama, S Hayashi, M Takahashi, and Y Hayashi. 1995. "Urinary bladder carcinogenesis induced by melamine in F344 male rats: correlation between carcinogenicity and urolith formation." *Carcinogenesis* 16(11):2773-7.

Panesar, NS, KW Chan, WS Lo, VH Leung, and CS Ho. 2010. "Co-contamination, but not mammalian cell conversion of melamine to cyanuric acid the likely cause of melamine-cyanurate nephrolithiasis." *Clin Chim Acta* 411(21-22):1830-1.

Partanen, H, K Vähäkangas, CS Woo, S Auriola, J Veid, Y Chen, P Myllynen, and H El Nezami. 2012. "Transplacental transfer of melamine." *Placenta* 33(1):60-6.

Phromkunthong, W, N Nuntapong, M Boonyaratpalin, and V Kiron. 2013. "Toxicity of melamine, an adulterant in fish feeds: experimental assessment of its effects on tilapia." *J Fish Dis* 36(6):555-68.

Puschner, B, RH Poppenga, LJ Lowenstine, MS Filigenzi, and PA Pesavento. 2007. "Assessment of melamine and cyanuric acid toxicity in cats." *J Vet Diagn Invest* 19(6):616-24.

Reimschuessel, R, E Evans, WC Andersen, SB Turnipseed, CM Karbiwnyk, TD Mayer, C Nochetto, NG Rummel, and CM Gieseker. 2010. "Residue depletion of melamine and cyanuric acid in catfish and rainbow trout following oral administration." *J Vet Pharmacol Ther* 33(2):172-82.

Rodriguez, PN, and AS Klein. 1988. "Management of urolithiasis during pregnancy." *Surg Gynecol Obstet* 166(2):103-6.

Root, DS, T Hongtrakul, and WC Dauterman. 1996. "Studies on the Absorption, Residues and Metabolism of Cyromazine in Tomatoes." *Pestic Sci* 48, 25-30.

Schell-Feith, EA, JE Kist-van Holthe, and AJ van der Heijden. 2010. "Nephrocalcinosis in preterm neonates." *Pediatr Nephrol* 25(2):221-30.

Shang, P, H Chang, ZJ Yue, W Shi, H Zhang, X Tang, Q He, and W Wang. 2012. "Acute kidney injury caused by consumption of melamine-contaminated infant formula in 47 children: a multi-institutional experience in diagnosis, treatment and follow-up." *Urol Res* 40(4):293-8.

Shen, JS, JQ Wang, HY Wei, DP Bu, P Sun, and LY Zhou. 2010. "Transfer efficiency of melamine from feed to milk in lactating dairy cows fed with different doses of melamine." *J Dairy Sci* 93(5):2060-6.

Shi, GQ, ZJ Wang, ZJ Feng, YJ Gao, J Di Liu, T Shen, M Li, et al. 2012. "A survey of urolithiasis in young children fed infant formula contaminated with melamine in two townships of Gansu, China." *Biomed Environ Sci* 25(2):149-55.

Stine, CB, R Reimschuessel, Z Keltner, CB Nochetto, T Black, N Olejnik, M Scott, et al. 2014. "Reproductive toxicity in rats with crystal nephropathy following high doses of oral melamine or cyanuric acid." *Food Chem Toxicol* 68:142-53.

Suchý, P, P Novák, D Zapletal, and E Straková. 2014. "Effect of melamine-contaminated diet on tissue distribution of melamine and cyanuric acid, blood variables, and egg quality in laying hens." *Br Poult Sci* 55(3):375-9.

Sugita, T, H Ishiwata, and K Yoshihira. 1990. "Release of formaldehyde and melamine from tableware made of melamine-formaldehyde resin." *Food Addit Contam* 7(1):21-7.

Sun, H, K Wang, H Wei, Z Li, and H Zhao. 2016. "Cytotoxicity, organ distribution and morphological effects of melamine and cyanuric acid in rats." *Toxicol Mech Methods* 26(7):501-510.

Sun, N, Y Shen, Q Sun, XR Li, LQ Jia, GJ Zhang, WP Zhang, et al. 2009. "Diagnosis and treatment of melamine-associated urinary calculus complicated with acute renal failure in infants and young children." *Chin Med J (Engl)* 122(3):245-51.

Sun, P, JQ Wang, JS Shen, and HY Wei. 2011. "Residues of melamine and cyanuric acid in milk and tissues of dairy cows fed different doses of melamine." *J Dairy Sci* 94(7):3575-82.

Sun, P, JQ Wang, JS Shen, and HY Wei. 2012. "Pathway for the elimination of melamine in lactating dairy cows." *J Dairy Sci* 95(1):266-71.

Takagi, K, K Fujii, K Yamazaki, N Harada, and A Iwasaki. 2012. "Biodegradation of melamine and its hydroxy derivatives by a bacterial consortium containing a novel Nocardioides species." *Appl Microbiol Biotechnol* 94(6):1647-56.

Tufro-McReddie, A, LM Romano, JM Harris, L Ferder, and RA Gomez. 1995. "Angiotensin II regulates nephrogenesis and renal vascular development." *Am J Physiol* 269(1 Pt 2):F110-5.

USEPA. n. d. *Assessing and Managing Chemicals under TSCA*. https://www.epa.gov/assessing-and-managing-chemicals-under-tsca.

Vanachayangkul, P, and WH Tolleson. 2012. "Inhibition of heme peroxidases by melamine." *Enzyme Res* 2012:416062.

Wackett, LP, MJ Sadowsky, B Martinez, and N Shapir. 2002. "Biodegradation of atrazine and related s-triazine compounds: from enzymes to field studies." *Appl Microbiol Biotechnol* 58(1):39-45.

Walker, KA, X Cai, G Caruana, MC Thomas, JF Bertram, and MM Kett. 2012. "High nephron endowment protects against salt-induced hypertension." *Am J Physiol Renal Physiol* 303(2):F253-8.

Wang, HY, H Wang, LL Tang, YH Dong, L Zhao, and G Toor. 2014. "Sorption characteristics of cyromazine and its metabolite melamine in typical agricultural soils of China." *Environ Sci Pollut Res Int* 21(2): 979-85.

Wang, IJ, YN Wu, WC Wu, G Leonardi, YJ Sung, TJ Lin, CL Wang, et al. 2009. "The association of clinical findings and exposure profiles with melamine associated nephrolithiasis." *Arch Dis Child* 94(11):883-7.

Wang, W, H Chen, B Yu, X Mao, and D Chen. 2014. "Tissue deposition and residue depletion of melamine in fattening pigs following oral administration." *Food Addit Contam Part A Chem Anal Control Expo Risk Assess* 31(1):7-14.

Wen, JG, QL Chang, AF Lou, ZZ Li, S Lu, Y Wang, YL Wang, et al. 2011. "Melamine-related urinary stones in 195 infants and young children: clinical features within 2 years of follow-up." *Urol Int* 87(4):429-33.

WHO. 2012. *UN strengthens regulations on melamine, seafood, melons, dried figs and labelling.* Jul 4. http://www.who.int/mediacentre/news/releases/2012/codex_20120704/en/index.html.

Wu, YT, CM Huang, CC Lin, WA Ho, LC Lin, TF Chiu, DC Tarng, CH Lin, and TH Tsai. 2009. "Determination of melamine in rat plasma, liver, kidney, spleen, bladder and brain by liquid chromatography-tandem mass spectrometry." *J Chromatogr A* 1216(44):7595-601.

Yan, H, J Wu, G Dai, A Zhong, J Yang, H Liang, and F Pan. 2010. "Interaction between melamine and bovine serum albumin: spectroscopic approach and density functional theory." *J Mol Struct* 967(1-3):61-4.

Yang, H, Q Wang, J Luo, Q Li, L Wang, CC Li, G Zhang, Z Xu, H Tao, and Z Fan. 2010. "Ultrasound of urinary system and urinary screening in 14

256 asymptomatic children in China." *Nephrology (Carlton)* 15(3):362-7.

Yang, J, L An, Y Yao, Z Yang, and T Zhang. 2011. "Melamine impairs spatial cognition and hippocampal synaptic plasticity by presynaptic inhibition of glutamatergic transmission in infant rats." *Toxicology* 289(2-3):167-74.

Yang, JJ, YT Tian, Z Yang, and T Zhang. 2010. "Effect of melamine on potassium currents in rat hippocampal CA1 neurons." *Toxicol In Vitro* 24(2):397-403.

Yang, Y, GJ Xiong, DF Yu, J Cao, LP Wang, L Xu, and RR Mao. 2012. "Acute low-dose melamine affects hippocampal synaptic plasticity and behavior in rats." *Toxicol Lett* 214(1):63-8.

Yang, ZH, CM Zhang, T Liu, XF Lou, ZJ Chen, and S Ye. 2010. "Continuous renal replacement therapy for patients with acute kidney injury caused by melamine-related urolithiasis." *World J Pediatr* 6(2):158-62.

Yin, RH, C Huang, J Yuan, W Li, RL Yin, HS Li, Q Dong, XT Li, and WL Bai. 2019. "iTRAQ-based proteomics analysis reveals the deregulated proteins related to liver toxicity induced by melamine with or without cyanuric acid in mice." *Ecotoxicol Environ Saf* 174:618-629.

Yokley, RA, LC Mayer, R Rezaaiyan, ME Manuli, and MW Cheung. 2000. "Analytical method for the determination of cyromazine and melamine residues in soil using LC-UV and GC-MSD." *J Agric Food Chem* 48(8):3352-8.

Zhang, BK, YM Guo, and L Wang. 2012. "Melamine residues in eggs of quails fed on diets containing different levels of melamine." *Br Poult Sci* 53(1):66-70.

Zhang, L, LL Wu, YP Wang, AM Liu, CC Zou, and ZY Zhao. 2009. "Melamine-contaminated milk products induced urinary tract calculi in children." *World J Pediatr* 5(1):31-5.

Zheng, X, A Zhao, G Xie, Y Chi, L Zhao, H Li, C Wang, et al. 2013. "Melamine-induced renal toxicity is mediated by the gut microbiota." *Sci Transl Med* 5(172):172ra22.

Zhu, H, and K Kannan. 2019. "Melamine and cyanuric acid in foodstuffs from the United States and their implications for human exposure." *Environ Int* Epub 2019 Jun 25.

Zou, CC, XY Chen, ZY Zhao, WF Zhang, Q Shu, JH Wang, L Zhang, SJ Huang, and LL Yang. 2013. "utcome of children with melamine-induced urolithiasis: results of a two-year follow-up." *Clin Toxicol (Phila)* 51(6):473-9.

In: An Introduction to Melamine
Editor: Ashley Harris

ISBN: 978-1-53617-136-5
© 2020 Nova Science Publishers, Inc.

Chapter 3

USE OF MELAMINE-FORMALDEHYDE RESIN AS SHELL MATERIAL FOR MICROENCAPSULATION

María de la Paz Miguel[*]

Institute of Materials Science and Technology (INTEMA), University of Mar del Plata, CONICET, Mar del Plata, Argentina

ABSTRACT

Polymers are commonly used in the fabrication of protective coatings for metallic substrates. Research on self-repairing coatings based on the incorporation of microcapsules or nanocapsules loaded with corrosion inhibitors or film-former healing agents continues to increase because of potential economic benefits of this technology. This chapter deals with the use of melamine-formaldehyde resins as shell material for microcapsules. The microencapsulation of linseed oil as repairing agent was carried out via the in-situ emulsion polymerization method. Chemical composition of microcapsules is one of the key factors for keeping their physical integrity and stability in the course of their synthesis process, storage and handling

[*] Corresponding Author's Email:maria.miguel@fi.mdp.edu.ar.

operations to fabricate a coating. In this work, the melamine-formaldehyde resin was chosen to form the shell material due to its good thermo-mechanical and water-resistant properties. A comparison between the syntheses of microcapsules based on melamine-formaldehyde and urea-formaldehyde resins is included. Selection of emulsifiers is another key factor because the microcapsules formation is influenced by the emulsion stability. Therefore, a series of experiments was conducted changing the emulsifying system and a discussion of results is provided. The synthesis products were examined by optical microscopy and scanning electronic microscopy (SEM). Further characterization applied to a powder of microcapsules allowed to verify the effective repairing agent encapsulation.

Keywords: melamine-formaldehyde resin, microcapsules, self-healing, smart coatings, linseed oil

INTRODUCTION

Coatings are applied to the surface of objects for decorative or functional purposes, or both. There is a growing interest in smart coatings based on materials with special properties. These coatings allow spontaneous respond to external stimuli and provide protection against unwanted impacts. Smart coatings comprise self-cleaning, antifouling, flame-retardant and self-healing coatings, among others. Inspired by the capacity owned by living organisms, self-healing coatings are able to recover autonomously or with external aid their initial properties. There are different approaches to impart self-healing functionalities to coatings, such as the use of self-healing materials within the matrix, vascular methods and the incorporation of microcapsules.

A microcapsule is a small reservoir formed by a core made of active compound surrounded by a shell or wall, usually made of an organic polymer. In addition, there are also poly-nuclear core and multi-shell microcapsules. The shell isolates the active compound from the surrounding matrix, avoiding undesirable reactions and possible leakages. Microencapsulation applications are widespread nowadays, including the pharmaceutical, cosmetic, nutritional, agriculture, textile and material research fields. The diameter of commercial microcapsules may vary from few to several tens of microns.

Polymeric coatings are typically used to provide protection of metallic parts from the corrosive environment by creating a barrier that delays the diffusion of chemical species to and from the metal surface. Because polymeric coatings could suffer microcracks and lose their properties, the embedment of self-healing microcapsules into the coatings has been proposed to address this problem. This approach would allow to prolong their lifetime and improve their anticorrosion performance. The phenomenon of self-healing takes place when the coating suffers microcracks and microcapsules containing a polymerizable material break releasing the repairing fluid, which fills the discontinuities and fixes the damaged matrix.

One of the most popular processes used to prepare microcapsules with a core-shell structure is by in situ polymerization of urea-formaldehyde (UF) or melamine-formaldehyde (MF) as shell material. The process involves the preparation of a stable oil-in-water emulsion in the presence of emulsifiers, followed by the oligomers polymerization and chain cross-linking to form a shell around the dispersed tiny oil drops. The first step is the nucleophilic addition reaction to obtain the prepolymer, which is catalyzed both by acids and bases. The second step is the condensation reaction, which is mainly acid catalyzed. Then, it is possible to produce core-shell structures by two alternative methods. The two-step method consists in a first step of prepolymer preparation in basic medium at about 70°C, followed by a condensation and encapsulation step in acidic medium at about 60°C. While in the one-step method, prepolymer preparation, condensation and encapsulation are carried out in acidic medium.

The two-step method is currently chosen to produce MF-based microcapsules (MC) because the prepolymer dissolves easily in an alkaline medium. The characteristic reactions are shown in Figure 1, denoting a melanine ring by ϕ. In fact, when melamine reacts with formaldehyde a family of nine methylolamines can be obtained by replacement of hydrogens in melamine by methylol groups. During the condensation step, two mechanisms are involved in resin formation: creation of ether bridges through the reaction between two methylol groups, and creation of methylene bridges trough the reaction between an amino group and a methylol group with elimination of water [1, 2]. The reaction of MF monomers and prepolymer in acidic aqueous phase leads to low molecular weight oligomers. Yuan *et al.* reported that the presence of hydrophobic groups and hydrophilic groups in prepolymer and oligomers molecules gives them a surfactant-like surface activity, which would drive their molecules towards the oil/water phase boundary. Therefore, concentration of reactive resin molecules at the phase boundary would favor resin condensation take place there faster than in the homogenous phase. The authors explained that emulsifier molecules tend to place hydrophobic groups orientated into the oil droplets and hydrophilic groups out of the oil droplets, forming a steric boundary layer around each droplet. That layer is permeable to MF prepolymer and oligomers and allows them to accumulate at the interface due to their surface activity. As a result, gel-like structures are built around oil drops, which further harden by cross-linking forming the MC shells [3].

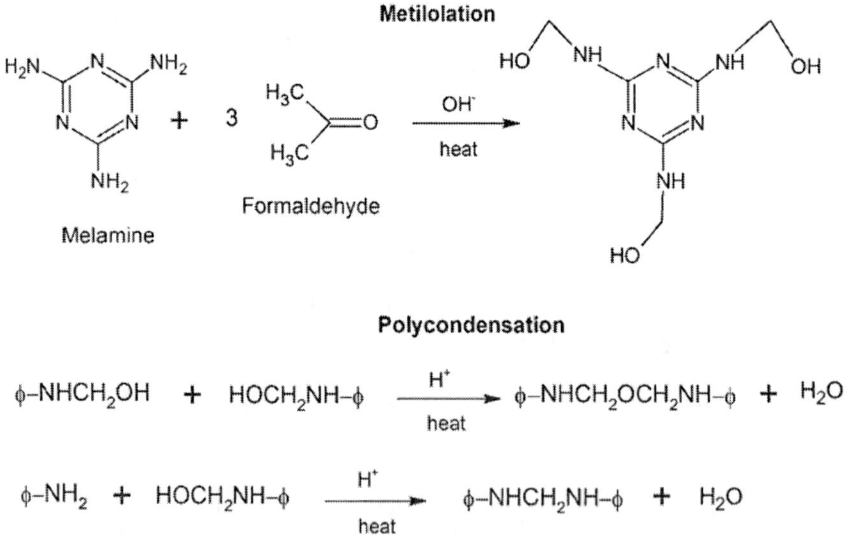

Figure 1. Reaction schemes of formation of prepolymer and MF resin.

MF and melamine-urea-formaldehyde (MUF) resins are usually preferred to UF resin due to a higher resistance to water attack. Melamine (2,4,6-triamino-1,3,5-triazine) is a very reactive monomer with three amino groups, which react with electrophiles, leading to highly cross-linked networks. Polycondensation reactions of melamine with formaldehyde yield the MF resins. MF resins have been widely used in coatings, tableware, furniture and adhesives and are recently being studied as a suitable material for the preparation of porous supports for selective trapping [2, 3, 4, 5]. That broad spectrum of applications is due to their outstanding properties, such as toughness, chemical resistance, thermal stability and low cost, which has also prompted a different application that will be explored along this chapter, the use of melamine-.formaldehyde resin as shell material for microencapsulation.

Various core substances have been encapsulated with MF resin such as phase change materials, essence oils, pesticides, flame retardants, polimerizable species for self-healing materials, photochromic compounds, among others [6-26]. Along this chapter, relevant findings arose from several investigations are cited to demonstrate the versatility of MF resin for microencapsulation applications. We include results of various syntheses of MF-based MC loaded with linseed oil (LO), also known as flaxseed oil or flax oil, performed with the aid of different combinations of emulsifiers. The aim has been to produce a dry powder of MC loaded with linseed oil. LO has the ability to polymerize in presence of oxygen air. Therefore, its encapsulation could be useful for the development of anticorrosive coatings by self-healing phenomenon [27-37].

The chapter is divided in five parts, where the use of MF resin as shell material for microencapsulation is described for different applications. Findings related to the influence of experimental conditions on the quality of MF-based microcapsules previously reported were included. Particularly, part number four deals the synthesis of MF-based microcapsules filled with linseed oil. Analysis about the effect of the emulsifying/stabilizing system on the performance of different syntheses was included. The chapter contains initial results about the preparation of smaller capsules with the aid of an ultrasonic processor. Finally, one part was devoted to recognizing possible changes in market trends regarding the use of UF and MF resins in daily life products, such as coatings and cleaning products [40].

1. APPLICATIONS

1.1. Microencapsulation of Phase Change Materials and Flame Retardants with MF Resin

Phase change materials (PCMs) are capable of absorbing, storing and releasing large amounts of energy as latent heat over a defined temperature range during their phase transitions. Their microencapsulation overcame many drawbacks when they are used in bulk, such as super-cooling, swelling and fluidity. Microcapsules (MC) with PCMs have the ability of phase change at melting points as surrounding temperature changes. Due to their thermoregulation or thermosaving abilities, they have been studied as coolants, solar and nuclear heat storage systems, packed bed heat exchangers, and they can be used for the manufacture of thermal regulated textiles, coatings and foams. Su *et al.* synthetized MC filled with dodecanol as phase change material. Methanol-modified melamine-formaldehyde was used as shell material for MC and styrene maleic anhydride (SMA) copolymer was used as dispersant [6]. Yin *et al.* synthetized MC with high thermal energy storage density using MF resin as shell material and n-hexadecanol as core material. They used SMA copolymer as an emulsifier. They explained that negatively charged copolymer molecules self-assemble on the surface of n-hexadecanol droplets and facilitate the precipitation of positively charged MF prepolymer onto the droplet surface electrostatically [7]. Alič *et al.* studied the microencapsulation of butyl stearate as phase change material with MF resin. They reported how different decreasing of pH regimes during the microencapsulation process influences on the composition, morphology and thermal stability of MC. They observed that a decrease to a low pH in a single step resulted in a rapid increase of the weight percentage of MF resin in samples and it led to MC with poor thermal stability and unsatisfactory morphology. They recommended decreasing the pH value in several small steps or in a linear way during microencapsulation. It would ensure slow curing reaction and deposition of MF resin on the MC surface, enabling to obtain MC with thicker shell and better thermal stability [8]. Konuclu *et al.* investigated the encapsulation of decanoic acid through the one-step method with UF, MF and MUF resins for thermal energy storage applications. On one hand, it was concluded that decanoic acid-filled MC with UF-based shell could achieve a good thermal energy storage capacity, but they were fragile. On the other hand, MC with MF-based shell offered better heat resistance, but they had lower thermal energy story capacity than the former ones. They recommended the use of MUF resin to encapsulate decanoic acid for thermally stable and leakage-free applications above 95°C, in the presence of the binary surfactant system formed by Tween 40 and Tween 80 [9]. Zhang and Wang developed a series of microencapsulated phase change materials, based on n-octadecane as core material and MF resin as shell material, through the two-step method. They obtained MC with regular spherical by using SMA as anionic emulsifier. It was observed that MC prepared with the aid of another anionic surfactant, sodium dodecylsulfate (SDS), displayed a poor regularity in shape and owned a thick shell. The use of poly(vinyl alcohol) (PVA) as a nonionic emulsifier, led to agglomeration of irregular in shape MC [10]. Wang and Zhao evaluated the influence of pH in the shell formation process during the encapsulation of n-octadecane via the two-step method. They suggested using styrene-maleic SMA as an emulsifier and performing the condensation step at a moderate pH value, around 4.6, to obtain regular spherical capsules with high encapsulation efficiency [11]. Wang *et al.* studied the microencapsulation of n-octadecane with MF resin shell by using SDS, Span-80 and Tween-80 as dispersants [12].

Flame retardants (FR) are chemical compounds incorporated to thermoplastics and thermosets polymers to provide varying degrees of flammability protection. Luo *et al.* synthesized MC containing decabromodiphenyl ether with MF resin as shell material by the two-step method, as a strategy to improve flame retardant behavior of thermoplastic resins [13]. Recently, phosphorous–containing flame retardants have been developed to replace halogen-containing flame retardants. Microencapsulation with MF resin of phosphorous-containing flame retardants allows to overcome the disadvantages of poor resistance to water attack and low compatibility with polymers. Wu *et al.* encapsulated aluminum hypophosphite with MF resin. The prepared MF-based MC allowed to increase the water resistance of flame-retardant acrylonitrile-butadiene-styrene composites after hot wat treatment [14]. Du *et al.* investigated the nanoencapsulation of n-octadecane with phosphorus-nitrogen containing diamine- modified MF resin as shell, by the two-stage method with the aid of SMA as an emulsifier. In the authors point of view, the phase change and flame-retardant material would be advantageous for energy saving and thermal energy storage applications [15].

1.2. Microencapsulation of Volatile Compounds

Essence oils are aromatic substances obtained from a botanically defined plant raw material. They are unstable and fragile volatile compounds. They are used for their aroma, flavor, biocidal activities and medicinal properties [41, 42, 43]. They can be encapsulated to improve their efficacy, protect them from environmental degradation, and facilitate their handling through solidification of the liquid oils. Lee *et al.* investigated the microencapsulation of floral essence oil with MF resin by the two-step method, using a mixture of sodium dodecylsulfate (SDS) and Tween 20 as emulsifiers and PVA as stabilizer. They observed that as formaldehyde-melamine molar ratio (F/M) increased, content of the residual formaldehyde in the preparation of prepolymer increased. The authors found that, at a relatively high pH value of reaction medium, a smooth MF-based shell was obtained, while at a low pH value, the surface morphology of MC was raspberry-like. This was explained by the adsorption of aggregates or premature particles onto the oil droplets [17]. J Hwang *et al.* encapsulated peppermint oil with MF resin in the presence of different emulsifiers. They observed extreme agglomeration with the use of gum arabic (GA). MC with a rough surface could be obtained in the presence of SDS, while MC with a smooth surface could be prepared with the aid of Tween 20. Before the cross-linking reaction, they added poly(vinyl alcohol) as a protective colloid to enhance the stability of MC [18]. Navarro *et al.* synthesized MF-based nanocapsules (NC) loaded with essence oils as natural biocides for footwear applications, using SDS as emulsifier [19]. Fei *et al.* also encapsulated essence oil but they used a methanol-modified melamine-formaldehyde resin as shell material. They used styrene-maleic anhydride copolymer as anionic dispersant and studied the effect of regulating the pH value at different times during the synthesis process. According to that research, in the case of pH regulation after prepolymer addition, small MC with a wide size distribution would be synthesized. There would be a slight increase in viscosity, which would enhance coacervation and formation of small MC. However, an increase in prepolymer reactivity could make some vesicles be cured before redispersion, leading to a larger size distribution. In turn, if the pH was fixed after heating, viscosity would rise slightly and reactivity would increase greatly. Therefore, large MC with a broad size distribution would be produced [20]. Due to the ability of MF resin to form a high densely packed and cross-linked matrix, it is widely used for the microencapsulation of volatile compounds in laundry-type applications. In that application, microcapsules are expected to bear exposure to detergents and washing cycle, and release encapsulated fragrances on fabric for prolonged periods [40].

1.3. Microencapsulation of Self-Healing Materials with MF Resin

Microcapsules loaded with self-healing agents are incorporated to polymeric matrixes to impart autonomic self-healing capability. Upon damage of the composites, healing agent is released from the broken MC and delivered to the separate portions, bonding the cracks through chemical reaction and/or physical interaction. Ming Meng *et al.* encapsulated glycidyl methacrylate (GMA) with MF resin for making self-healing epoxy materials. They used sodium dodecyl benzenesulfonate (SDBS) as emulsifier and PVA to protect the colloid and prevent the newly formed MC from sticking [23]. Yuan *et al.* prepared microcapsules containing curing agent for epoxy. MF resin was used as shell material and high-activity polythiol (pentaerythritol tetrakis (3-mercaptopropionate), PETMP) was used as core substance. Emulsion was prepared in the presence of poly(styrene-maleic sodium) solution as emulsifier by using a homogenizer for 5 minutes at a selected rate, varying from 10000 to 20000 rpm. The authors observed that a rise in the emulsion dispersion rate decreases the MC average diameter and their size distribution becomes narrower. In addition, when they increased emulsifier content, a finer emulsion could be obtained [3]. Zhu *et al.* prepared MF-based MC filled with bisphenol A type epoxy resin E-51 in the presence of different emulsifiers. They found that the use of SMA arouses emulsion breaking, while the use of the SDS and SDBS keeps the emulsion stability due to the tendency of their molecules to be adsorbed on the oil-in-water interface with intensive and highly orientated arrangement. They used PVA to reduce the oil drops size. They recommended to perform the condensation reaction at a pH value of 3.5-4 because higher pH values would lead to large-sized MC with rough and loose MF shell, and lower pH values would arouse a wide size distribution [24]. Sharma and Chouldary analyzed the effect of melamine-formaldehyde ratio on the shell stability of MF-based MC loaded with epoxy resin for the development of self-healing composites [25]. Khorasani *et al.* prepared MC filled with coconut oil-based alkyd resin using MUF resin as shell material. They used the one-stage in-situ polymerization method with SDBS as an emulsifier [26].

Vegetable drying oils are reasonably inexpensive and eco-friendly [34]. They can be encapsulated for the development of anticorrosive coatings with self-healing capability. Linseed oil encapsulation has aroused interest as a corrosion inhibitor agent, either alone or mixed with other species [27-37]. Its encapsulation with MF resin is studied in the following part of the text.

1.4. Encapsulation of Linseed Oil with MF Resin

1.4.1. Linseed Oil Encapsulation Background

Encapsulation of linseed oil has attracted attention for the development of anticorrosive coatings with self-healing capacity. The molecular structure of a representative triglyceride found in linseed oil is displayed in Figure 2. Linseed oil possess a high content of unsaturated fatty acids, which allows it to polymerize in presence of oxygen forming a stable waterproof film [37, 39]. Suryanarayana *et al.* encapsulated LO with cobalt naphthenate and lead octoate as driers with UF resin using PVA as emulsifier [27]. Selvakumar *et al.* synthesized MC with LO which contained driers and nanoparticles of CeO_2 and Cr_2O_3 as corrosion inhibitors [28]. Hasanzadeth *et al.* synthesized UF-based MC with LO as the healing agent with and without nanoparticles of CeO_2 [29]. S. Hatami Boura *et al.* encapsulated LO in presence of PVA [30]. Nesterova *et al.* prepared LO-filled MC following the method reported by Suryanarayana *et al.* [31]. Lang *et al.* discussed the effect of the use of different molecular weight PVA on the preparation of UF-based MC with LO [32]. Wang *et al.* studied the anticorrosion behavior of a scratched epoxy coating with embedded UF-based MC filled with linseed oil [33]. Behzadnasab *et al.* studied the corrosion performance of a self-healing epoxy-based coating containing LO-filled MC and suggested the incorporation of active inhibitors to LO in order to extend the barrier anticorrosive effect in the final product [34]. Linseed oil was encapsulated with MUF resin in presence of poly(vinylpyrrolidone) as emulsifier and colloid protector by Khalaj Asadi *et al.* The authors observed that the microencapsulation yield improved, the mean particle size decreased and the outer surface of MC became smoother as PVP molecular weight increased [35]. Abdipour *et al.* prepared UF-based capsules containing LO in the presence of a solution of PVA and SDS as emulsifying/stabilizing system. The prepared capsules were separated from the suspension by vacuum filtration. After washing the MC with distilled water and xylene, the authors store the MC in a solution of distilled water and xylene. The capsules could endure even a few months of storage before being incorporated in a epoxy coating [36]. Typically, UF resin has been used as shell material for LO encapsulation studies. The preparation of UF-based MC loaded with LO following the one-stage and two-stage methods has been studied by our research group [44].

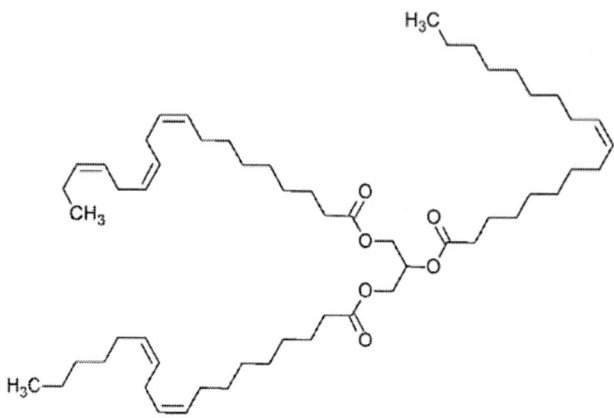

Figure 2. Molecular structure of a representative triglyceride found in linseed oil.

1.4.2. Synthesis of MF-Based MC Loaded with Linseed Oil

In this section, focus was placed on the synthesis of MF-based MC loaded with linseed oil. Results of several syntheses performed with different combinations of emulsifiers are included. The aim was to produce a dry powder of LO-filled MC with a MF-based robust shell, so that it could eventually be incorporated into a host resin matrix. A selection of results has been made, although the complete study can be found in reference [45].

1.4.2.1. Materials Used for the Studied Syntheses

Shell materials: Melamine p. a., formaldehyde (37wt.% aqueous solution), ammonium chloride (≥99.5 wt.%) and resorcinol (>99 wt.%). Core material: Linseed oil (commercial grade). Emulsifiers/Stabilizers: Sodium dodecylsulfate (SDS), polyvinyl alcohol (PVA) (Mw: 10000-30000 g/mol), sodium dodecylbencensulfonate (SDBS), purified gum arabic in powder, unpurified gum arabic in grains. Antifoaming agent: 1-octadecanol (>99 wt.%). Agents for pH tuning: Sodium hydroxide (≥97 wt.%), glacial acetic acid (≥99 wt.%), triethanolamine (TEA) (≥99 wt.%). Solvents: Commercial medicinal ethanol, double distilled water and xylene.

1.4.2.2. Procedure to Obtain MF-Based Shell by the Two-Step Method

The first step consists in the preparation of prepolymer: 1.81 g melamine, 3.5 g formaldehyde (37 wt.%), and 5 g of distilled water were added to a vial. Then pH was adjusted to 8-9 with 2-3 droplets of TEA. The mixture was magnetically stirred at 200 rpm over a hotplate at 70°C and left to react for an hour. The second step bases on performing the condensation process: The prepolymer solution was transferred to a 250 ml beaker with 80 g of distilled water and the mixture was mechanically agitated at a 600 rpm stirring rate. The beaker was set in a thermostatic bath, which initially remained at 35°C. Then, a mixture of 0.35 g NH_4Cl, 0.18 g resorcinol and 10 g of distilled water was added to the system. After isolating the system, temperature was gradually increased to 60°C, while pH was decreased to four. Afterwards, the system was left to react under stirring for 1.5 hours at 60°C. Before stopping agitation, the system was cooled to room temperature. The pH was brought to 7-8 with aqueous sodium hydroxide solution (5 wt.%) to stop the condensation reaction, which is acid catalyzed. The synthesized resin particles stayed at the bottom of the vessel.

1.4.2.3. Procedure to Obtain MF-Based Microcapsules by the Two-Step Method

A chart with the amounts of materials used for each synthesis, according to the following procedure, is shown in Figure 3. The first step involves the preparation of prepolymer, as previously described, and the preparation of emulsion. To prepare the emulsion a certain amount of LO was poured into a beaker containing the selected emulsifier combination. A volume of 50 mL of aqueous solution was completed by addition of distilled water in case it was necessary. The pH was tuned to 8-9 with TEA. Emulsion of LO was obtained with the aid of a homogenizer, at a nominal speed of 11000 rpm for 10 minutes. The second step bases on performing the condensation and microencapsulation process. The emulsion was slowly put into a 250 mL beaker containing 50 g of distilled water. The beaker was immersed in a thermostatic bath, which was initially at 35 °C. The system was stirred at a 600 rpm. If necessary, a few drops of 1-octadecanol were added as antifoaming agent. Then, the prepolymer was poured into the vessel and, 10-15 minutes later, a mixture of NH₄Cl, resorcinol and 10g of distilled water was also added to the beaker. After isolating the vessel, temperature was progressively increased to 60°C while pH was slowly reduced to four during 2 hours. Next, the system was left to react under stirring for 1.5 hours at 60°C. After that, synthetized final mixture was cooled to room temperature. The pH was adjusted to 7-8 with aqueous sodium hydroxide solution (5 wt.%), and then agitation was stopped.

Parameter	Synthesis I	Synthesis II	Synthesis II	Synthesis IV
Melamine (g)	1.89	1.89	1.81	1.81
Formaldehyde (37 wt. %) (g)	3.72	3.72	3.5	3.5
Linseed oil (g)	13.18	13.18	10.9	10.9
PVA solution (10 wt. %) (mL)	10	20	-	-
SDBS solution (0.5 wt. %) (mL)	15	30	-	-
Commercial unpurified GA in grains solution (5 wt. %) (mL)	-	-	-	10
Purified GA in powder solution (5 wt. %) (mL)	-	-	10	-
SDS solution (5 wt. %)	-	-	10	10
NH₄Cl (g)	0.387	0.387	0.36	0.36
Resorcinol (g)	0.189	0.189	0.18	0.18

Figure 3. Materials used for each of the four MF-based MC syntheses: I, II, III and IV. Reprinted from Ref. [45], with permission of Elsevier.

The same formaldehyde-melamine molar ratio (3/1) was used for the four syntheses. The weight core-shell ratio for synthesis I and II was around 3.4, while for synthesis III and IV it was around 3.

1.4.3. Conditioning Operations Applied on Synthesis Products

Conditioning operations were applied on the synthesized final mixture with the aim to produce a powder of MC. It is worth to mention that shell toughness is key to avoid breakage during MC conditioning operations or subsequent mixing with the host resin matrix, storage and/or handling of the final composite.

Isolation of MC from reaction medium was performed by centrifugation with two parts of ethanol-water solution 1:1, at approximately 2000 rpm. As a result, three phases were fractioned: a top layer with microcapsules of greater size, an aqueous residual colloidal phase which was discarded, and a bottom layer which contained higher density smaller microcapsules and polymer particles. Synthesis products were dried in a stove at 32°C for 24 hours.

1.4.4. Characterization Methods

The presence of microcapsules and their size were observed by Transmission Optical Microscopy. Microcapsules surface was investigated by Scanning Electronic Microscopy (SEM). Microcapsules average diameter was determined through analysis of SEM images with the help of a specific software. Chemical characterization was analyzed by Confocal Raman Microscopy (CRM): Raman spectra of linseed oil, shell material and microcapsule were recorded on a Raman spectrometer coupled with a confocal microscope. Oil content was measured by extraction with xylene in a Soxhlet apparatus. Thermal stability of MC, LO and MF resin was analyzed by Thermo Gravimetric Analysis (TGA). Complementary characterization was performed by Scanning Differential Calorimetry (DSC). Results of thermal characterization can be found in Ref [45].

1.4.5. Morphological Characteristics of MF Resin

A SEM micrograph in Figure 4 of MF-based shell material reveals the presence of MF resin spherical particles a few microns in diameter. The resin was synthesized by the two-stage method. A broad particle size distribution is observed and the particle surface looks rough, probably due to the adsorption of tiny particles. The use of techniques to verify the effective encapsulation is certainly necessary to differentiate MC from polymer microparticles resulting from the procedure.

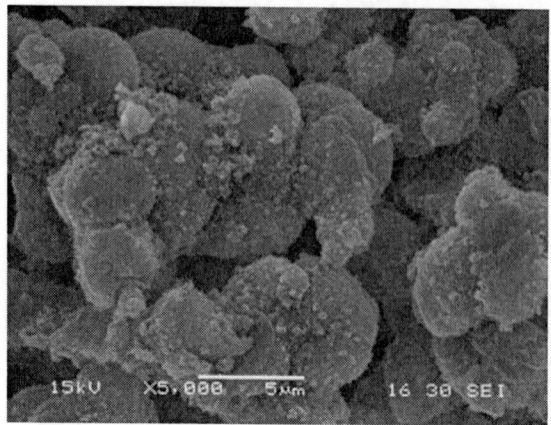

Figure 4. SEM micrograph of MF resin. Reprinted from Ref. [45], with permission of Elsevier.

1.4.6. Effect of the Emulsifying/Stabilizing System on Emulsion and MC Morphology

The purposes of using emulsifying/stabilizing system are to reduce interfacial tension between oil and aqueous phases allowing to obtain smaller MC and stabilize the emulsion, avoiding coalescence by its adsorption on the oil-water interface by forming a surrounding layer on oil droplets. Alic et al. suggested using anionic emulsifiers to produce MC with a homogenous and compact MF shell. They specified that the use of cationic emulsifiers is unadvisable to stimulate shell formation because of charge repulsion between cationic emulsifier and MF resin. H. Zhang and X. Wang [10] explained that MF oligomers molecules turn active in acidic solution by obtaining protons. Therefore, positively charged oligomers and prepolymer could be attracted by a negative electric field formed by anionic emulsifiers surrounding oil droplets. In addition, formation of hydrogen bonds between the hydroxyl groups in MF prepolymer and hydrophilic groups of anionic emulsifiers was also possible. The combination of two emulsifiers often exhibits better performance than the use of either emulsifier by itself, as long as it exits synergism in its interfacial properties. Thus, different binary mixtures of emulsifiers were tested for the encapsulation of LO.

Use of Melamine-Formaldehyde Resin as Shell Material ... 93

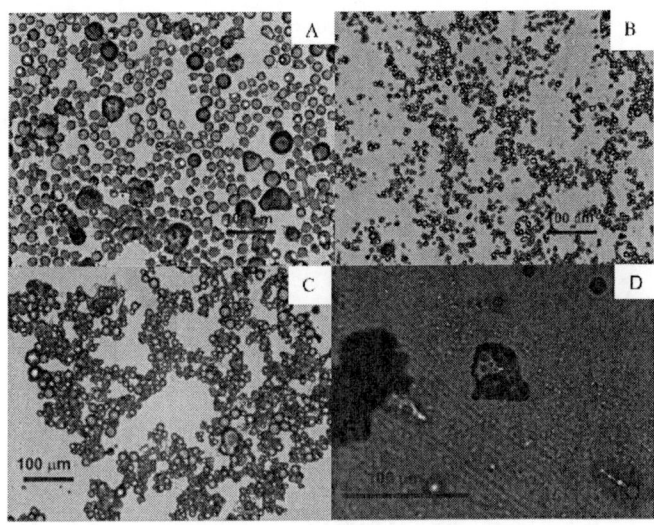

Figure 5. Optical micrograph (x10) of: A) top-layer sample of synthesis I. B) top-layer sample of synthesis II. C) Optical micrograph (x20) of top-layer sample of synthesis II, after drying. D) Optical micrograph (x50) of top-layer sample of synthesis III, after drying. Reprinted from Ref. [45], with permission of Elsevier.

As a result of a futile trial to produce a dry powder of UF-based MC loaded with LO in the presence of a combination of SDBS and PVA emulsifiers by using an homogenizer (T18 Ultra-turrax IKA unit), our research group wondered what happened whether resin was changed. Therefore, we tried to synthetize more resistant MF-based MC by using the same emulsifying/stabilizing system and agitation device. In addition, we performed the synthesis with the couples of emulsifiers formed by purified GA and SDS and unpurified GA and SDS. The effect of the emulsifying/stabilizing system was analyzed trough OM and SEM.

For each synthesis the top layer, containing essentially MC, was isolated from reaction medium as previously detailed by centrifugation. Figure 5.A shows a micrograph of a top-layer sample of synthesis I, where vesicles bigger than the microcapsules are observed. While Figure 5.B shows one of synthesis II, where tiny MC predominate. Unfortunately, for synthesis II, agglomeration was noticed after drying as observed in optical micrograph (x20) of Figure 5.C for the top-layer sample. Then, the MC obtained in synthesis II could not endure the drying process. In the case of top-layer sample of synthesis III, a marked agglomeration was observed after drying, as shown in Figure 5.D. Spherical MC can be clearly observed in optical micrograph of Figure 6, which belongs to a sample of top-layer material of synthesis IV. Fortunately, MC could endure the centrifugation and drying operations and a dry powder could be obtained.

Dissimilarity in chemical structure between commercial GA in grains and purified GA in powder is responsible for the difference among the products obained in the syntheses III and IV. Gum Arabic is a biopolymer chemically composed of carbohydrates and proteins. It is formed by a main fraction called Arabinogalactan (AG), a second fraction called Arabinogalactan-Protein complex (AGP) and a third small fraction of Glycoprotein (GP). It was reported earlier that the AGP complex is the main component responsible for GA emulsifying and stabilizing properties [46, 47, and 48]. Tabatabaee Amid *et al.* investigated the biopolymer obtained from durian seed and found that purification techniques reduced protein content and decreased the GA emulsifying capacity. The interfacial activity or emulsifying property would depend on the fraction of surface-active proteins attached to the polysaccharide [49].

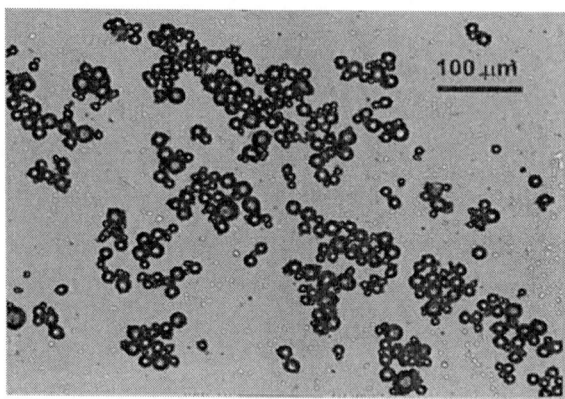

Figure 6. Optical micrograph (x10) of dried top- layer sample of synthesis IV. Reprinted from Ref. [45], with permission of Elsevier.

A SEM micrograph of a sample taken at the end of synthesis II before neutralization can be seen in Figure 7. It shows that by using 20 ml of PVA solution (10 wt.%) and 30 ml of SDBS (0.5 wt.%) as an emulsifying system, the obtained microcapsules were stuck and agglomerated even before neutralization. MC could have damaged during the drying of samples.

Crushing and adhesion of small particles against MC and smashed MC were observed in top-layer samples of synthesis III, as displayed in SEM micrograph of Figure 8.

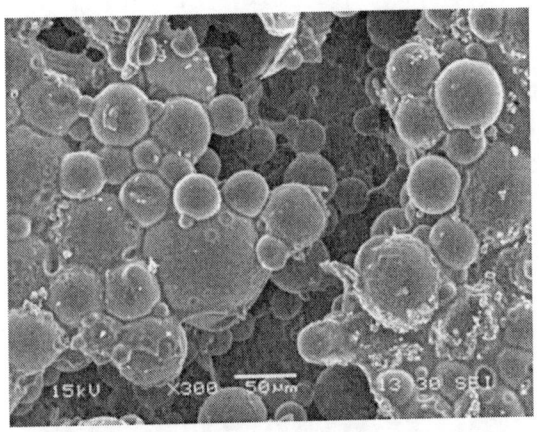

Figure 7. SEM micrographs for synthesis II of reaction medium sample before neutralization. Reprinted from Ref. [45], with permission of Elsevier.

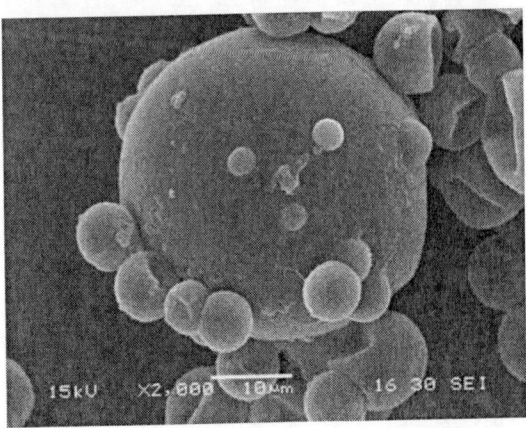

Figure 8. SEM micrograph for synthesis III of top-layer sample. Reprinted from Ref. [45], with permission of Elsevier.

SEM micrographs of top-layer sample of sinthesis IV reveal a large number of spherical microcapsules, as displayed in Figure 9.A. At first sight, the surface looks compact and smooth with some particles stuck on it, see Figures 9. B. A MC wall can be appreciated in Figure 9.C, with higher magnification; the surface formed by adhesion of tiny particles looks rather rough.

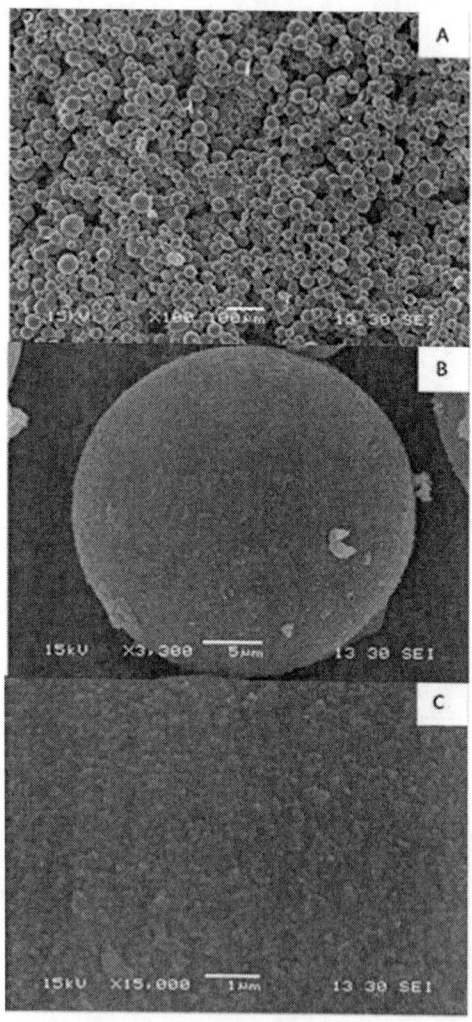

Figure 9. SEM micrographs for synthesis IV of: A) top-layer sample. B) MC of top-layer sample. C) MC wall. Reprinted from Ref. [45], with permission of Elsevier.

Further characterization was performed on the products of synthesis IV. SEM images of top-layer samples of synthesis IV were processed with image software to determine the particle size distribution and calculate the average diameter of MC. A wide range of MC diameters between 8 and 71 µm were measured, the estimated average diameter being 30.5 µm. The pH tuning after the addition of the prepolymer, while the reaction mixture was heated, led to a broad distribution of small MC in accordance with Fei et al.'s findings [20].

1.4.7. Chemical Characterization of MC

Raman spectroscopy is sensitive to molecular vibrational modes, particularly symmetric vibrations of non-polar groups and provides a fingerprint of a molecule [50]. CRM allowed to perform the chemical characterization of MC and verify core material encapsulation.

Raman spectra of the shell material, linseed oil and a top-layer microcapsule of synthesis IV can be seen in Figure 10. The corresponding to LO and top-layer MC sample were multiplied by a factor of three, in order to obtain better visualization to compare MC spectrum to MC main components spectra. The presence of peaks at 1269, 1443 and 1662 cm^{-1} in the Raman spectrum of LO was ascribed to cis-CH=CH wag, CH$_2$ bending and stretching of C=C group, respectively. A sharp peak at 978 cm^{-1} due to the triazine ring of melamine was observed at the Raman profile of the shell material sample. Other peaks in the 500-750 and 1200-1600 regions (cm^{-1}) were due to derivatives and substituted melamine. The similarity of MC Raman spectrum with that of LO over the measurement range allowed to prove the effective encapsulation of linseed oil with MF resin. The analysis of other samples threw similar results.

The characterization of oil content was performed by extraction with solvent. After extraction the solvent showed a yellowish color typical of linseed oil. A 78 wt.% of linseed oil content was determined by extracting oil with xylene.

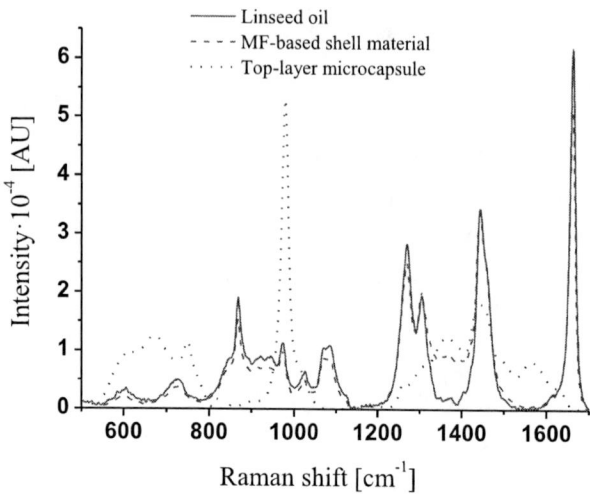

Figure 10. CRM spectra of MF-based shell material, pure linseed oil and top-layer microcapsule. Reprinted from Ref. [45], with permission of Elsevier.

1.4.8. Preparation of MF-Based Capsules with the Aid of an Ultrasonic Processor

The size distribution of the capsules depends mainly on the conditions used to prepare the emulsion. Different devices can be used for preparing the emulsion, such as magnetic stirrers, mechanical stirrers, dispersant homogenizers or ultrasonic processors. Agitation time and speed or power delivered by the device are relevant parameters. An increase in stirring rate can reduce drop size and make size distribution become narrower. However, the use of higher stirring speeds could favor droplets breakage, existing a practical maximum stirring speed. The kind and quantity of surfactants and/or stabilizers selected to prepare emulsion also have influence on the drop size.

In the present work, initial results about the synthesis of MF- based capsules loaded with linseed oil by means of an ultrasonic processor have been included.

Prepolymer was prepared, as previously described in part 4.2.2, by using 1.8 g of melamine, 3.5 g of formaldehyde and 5 g of distilled water. To prepare the emulsion 3.5 g of LO was poured into a beaker containing 25g of distilled. The pH was tuned to 8-9 with TEA. Emulsion of LO was obtained with the aid of an ultrasonic processor with microtip (Cole Palmer 750W). Synthesis A was carried out in a 100 mL beaker. The microtip was applied four times for three minutes at 20% of amplitude, spending 1730 J each time. Synthesis B was carried out in a centrifuge tube. The microtip was applied three times for three minutes at 20% of amplitude, spending 2200 J each time. At the sight of LO over the liquid surface, the amplitude was increased to 30% and the microtip was first applied during a minute, spending 1100J, and then four times for 3 minutes, spending 3365 J each time. The emulsion was slowly put into a 250 mL beaker containing 50 g of distilled water. and 20 mL of SDS solution (5 wt.%) The beaker was immersed in a thermostatic bath, which was initially at 35 °C. The system was stirred at a 600 rpm. After fifteen to twenty minutes of stabilization, the prepolymer was poured into the vessel. Ten minutes later, 10 mL of unpurified GA solution (20 wt.%) and a mixture of 0.36 g of NH_4Cl, 0.18 g of resorcinol and 10 g of distilled water were also added to the beaker. After isolating the vessel with aluminum foil, temperature was progressively increased to 60°C while pH was slowly reduced to 4.5 with acetic acid solution (5 wt.%) during 2 hours. Next, the system was left to react under stirring for 1.5 hours at 60°C. Before stopping agitation, synthetized final mixture was cooled to room temperature and neutralized with sodium hydroxide solution (5 wt.%).

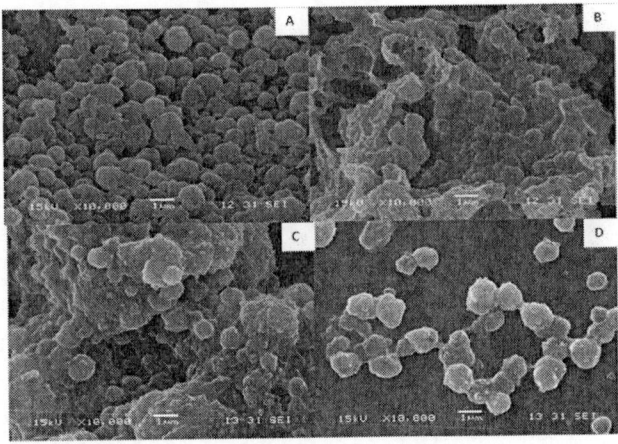

Figure 11. SEM micrographs for synthesis A (A and B) and synthesis B (C and D).

SEM micrographs of samples of both syntheses reveal the presence of agglomerated material, as displayed in Figures 11.A. and 11.B for synthesis A and in Figures 11.C and 11.D for synthesis B.

Results of exploratory work imply that tuning of synthesis parameters or changes in the experimental procedure need to be considered in greater depth to obtain a free-flowing powder of capsules.

1.4.9. Comparison with UF-Based Capsules

In this part, description and results from different syntheses of UF-based MC are included. Then, a comparison is made between UF-based and MF-based MC.

Materials: Shell materials: Urea p. a., formaldehyde (37 wt. % aqueous solution), ammonium chloride ($\geq$ 99.5 wt. %), resorcinol (> 99 wt. %). Core material: linseed oil (commercial grade). Emulsifiers/Stabilizers: Sodium dodecylbencensulfonate (SDBS) and polyvinyl alcohol (PVA) (Mw: 10000-30000 g/mol). Antifoaming agent: 1-octadecanol (> 99 wt. %). Agents for pH tuning: Hydrochloric acid (HCl) (37 wt. %), sodium hydroxide ($\geq$ 97 wt. %) and triethanolamine (TEA) ($\geq$ 99 wt. %).). Commercial medicinal ethanol was utilized to wash the microcapsules; and double distilled water (DW) was used to prepare the solutions and wash the microcapsules.

1.4.9.1. Preparation of UF-Based MC

A chart with the amounts of materials used for each synthesis, according to the following procedures, is shown in Figure 12.

1.4.9.1.1. Preparation of UF-Based MC by the Two-Step Method (Synthesis 1) by Mechanical Agitation

The first step consists in the preparation of prepolymer: urea, formaldehyde (37 wt. %), and 5 g of distilled water were added to a vial. Then pH was adjusted to 8-9 with 2-3 droplets of TEA. The mixture was magnetically stirred at 200 rpm over a hotplate at 70°C and left to react for an hour. To prepare the emulsion LO was poured into a 250 mL vessel containing the selected emulsifier combination. If necessary, a few drops of antifoaming agent were added to the system. The vessel was immersed in a thermostatic bath, which was initially at 35°C. The system was stirred at a 600 rpm and left to stabilize for 20 minutes. The second step bases on performing the condensation and microencapsulation process. The prepolymer was poured into the vessel and, 10-15 minutes later, a mixture of NH_4Cl, resorcinol and 20g of distilled water was also added to the vessel. After isolating the vessel with aluminum foil, temperature was progressively increased to 55°C while pH was slowly reduced to 3-4 with HCl solution (0.5 mol/L) during 2 hours. Then, the system was left to react under stirring for 1.5 hours at 55°C. After that, synthetized final mixture was cooled to room temperature. The system was neutralized with aqueous sodium hydroxide solution (5 wt. %), and then agitation was stopped.

1.4.9.1.2. Preparation of UF-Based MC by the One-Step Method (Syntheses: 2 and 3) with High-Performance Dispersing Devices

Urea was added to a 250 mL vessel containing 80g of distilled water. The vessel was immersed in a thermostatic bath, which was initially at 35°C. The system was agitated with a mechanical mixer at 600 rpm driving a four-bladed, 60 mm in diameter glasss propeller placed just above the bottom of the vessel. A mixture of NH_4Cl, resorcinol and 20g of distilled water was also added to the vessel. The pH value was tuned to 3-4 with HCl solution (0.5 mol/L). To prepare the emulsion, LO was poured to an 80 ml vessel containing the emulsifying combination. For synthesis 2, a homogenizer was immersed into the mixture at 11000 rpm for 10 min to form emulsion. While for synthesis 3, the probe of an ultrasonic processor was placed in the linseed oil-water mixture for 4 min at 40% intensity to form emulsion. After that, the system was left to stabilize for 20 minutes under stirring at a 600 rpm. Then formaldehyde (37 wt. %) was added to the system. After isolating the vessel with aluminun foil, temperature was progressively increased to 55°C. Next, the system was left to react under stirring for 4 hours at 55°C. Following that, final mixture was cooled to room temperature. The pH was adjusted to 7-8 with aqueous sodium hydroxide solution (5 wt. %), and then agitation was stopped.

Parameter	Synthesis 1	Synthesis 2	Synthesis 3
Urea (g)	2.8	2.5	2.5
Formaldehyde (37 wt. %) (g)	6.75	6.34	6.5
Linseed oil (g)	12.5	13.2	6
PVA solution (10 wt. %) (mL)	1.5	10	10
SDBS (0.5 wt. %)	100	15	15
NH_4Cl (g)	0.28	0.25	0.5
Resorcinol (g)	0.28	0.25	0.25

Figure 12. Materials used for each of the three UF-based MC syntheses: 1, 2, and 3.

1.4.9.2. Isolation of UF-Based MC from Reaction Mixture

Due to the density difference between LO (0.92 g/cm^3) and UF resin (1.15-1.19 g/cm^3), once the final reaction mixture of synthesis 1 was left to rest, the linseed-oil filled MC were at the top layer. Microcapsules were easily separated from the suspension by decantation. When Ultra-turax unit and ultrasonic homogenizer were used to prepare emulsions (Syntheses 2 and 3), decantation of MC was very slow. Thus, solids were separated by centrifugation at 500-2000 rpm. Then, solids were rinsed with distilled water or ethanol-water solution 1:1, vacuum-filtered and air-dried for 24-36 h at 22°C.

During microencapsulation two simultaneous reactions take place: the reaction of UF at LO-water interface to form the MC shell and the reaction of UF in solution to form colloidal particles. Few MC were obtained by the two-step method, and most of them were various tens of microns in diameter due to the low stirring speed used to form emulsion (600 rpm). The surface of MC was rough and porous due to the deposition and agglomeration of UF particles. It was observed that nanoparticles remained mainly in suspension, which could be attributable to a domine of the reaction leading to formation of UF colloidal particles; therefore, a strong shell was not formed. In addition, filtered solid was a sticky wet paste turning yellow after 1-2 days, indicating that the LO was not completely encapsulated. Figure 13 shows an optical micrograph of a sample taken from synthesis 1 reaction mixture at the end of reaction. Despite a group of MC can be seen, the material looks agglomerated. A SEM micrograph of a supernadant sample taken from synthesis 1 reaction mixture is displayed in Figure 14. The MC are surrounded by agglomerated material. Some MC look crushed.

When the microencapsulation procedure was carried out by the one-step method (syntheses 2 and 3) a great deal of MC were produced and the amount of UF particles in suspension was reduced. The surface of MC was rough and composed of UF nanoparticles potruding from the surface. Figure 15 shows a SEM micrograph from an aliquot taken from the top-layer obtained by centrifugation of synthesis 2 reaction mixture, where MC can be seen clearly. The average microcapsule diameter was around 50 μm. However, it was difficult to obtain a free-flowing power of MC because most of the times MC could not bear the conditioning operations. Conversely, when an ultrasonic homogenizer was used to form LO-water emulsion MC owned a rough and compact surface. The presence of smooth-surfaced capsules was also noticed, though they could not be clearly distinguished from UF colloidal particles formed in solution. MC prepared by synthesis 3 were strong enough to bear the preparation circumstances, and could be dried into a free-flowing powder. Figure 16 shows an optical micrograph of a sample taken of synthesis 3 solids. A great deal of small spherical MC can be observed. A deeper study can be found in reference [44].

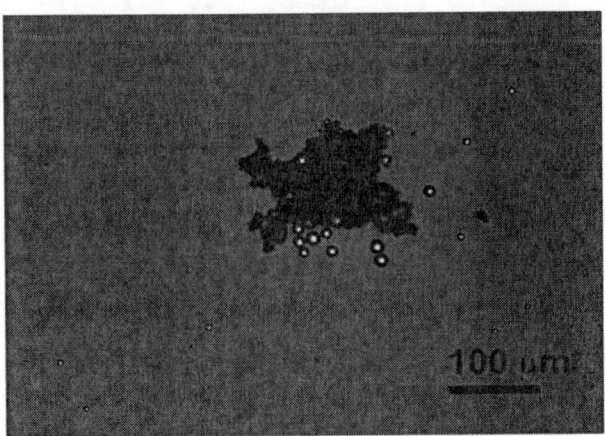

Figure 13. Optical micrograph (x20) of sample taken from synthesis 1 reaction mixture.

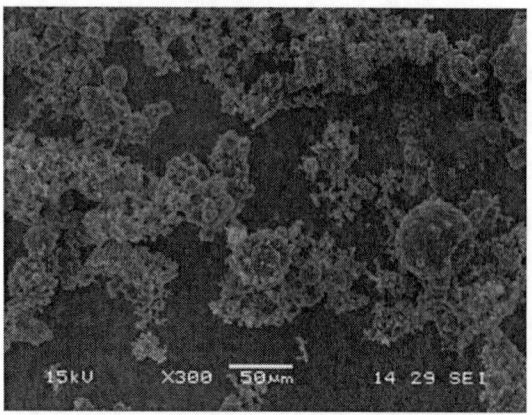

Figure 14. SEM micrograph of liquid sample of synthesis 1.

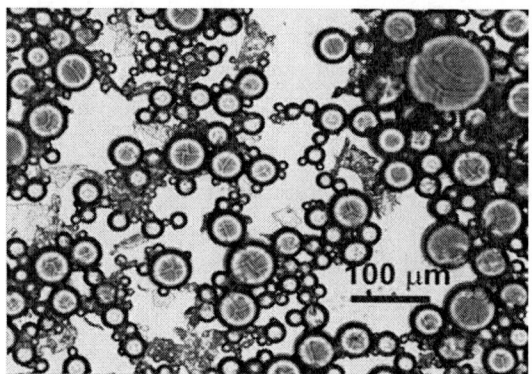

Figure 15. Optical micrograph (x20) of top-layer sample of synthesis 2.

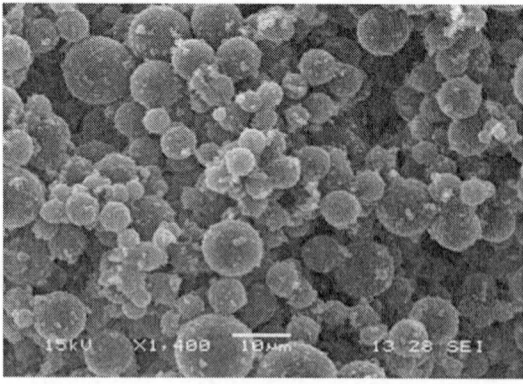

Figure 16. SEM micrograph of sample of synthesis 3.

1.4.9.3. Comparison of MC Prepared with UF Resin through the One-Step Method and MF-Based MC Prepared through the Two-Step Method

The synthesis of MF-based MC through the two-step method is more laborious than the synthesis of UF-based MC by the one-stage method because the first synthesis requires the prepolymer preparation and a slow tuning of pH value. When the Ultraturrax device was used to form the oil-in-water emulsion, a broad distribution of spherical microcapsules could be obtained through both syntheses. However, only MF-based MC prepared with the mixture of commercial unpurified GA in grain solution with SDS solution as emulsifiers could withstand the preparation and conditioning operations and allowed to obtain a free flowing powder. In that case, emulsion stabilization was achieved and the shell made of MF resin was more compact than the porous shell of MC prepared with UF resin in the presence of SDBS and PVA as emulsifiers. Therefore, UF-based MC were more fragile and broke releasing LO during conditioning operations. As a result of it, a sticky and agglomerated material was obtained.

Smaller UF-based MC prepared with the aid of an ultrasonic homogenizer were more resistant and could endure preparation and conditioning procedures and it was possible to produce a free-flowing powder of MC. In the case of preparation of MF-based capsules with the ultrasonic homogenizer, a deeper investigation is required to achieve emulsion stabilization and produce a powder of smaller MC.

1.5. Perspectives of the Use of UF and MF Resins as Shell Materials

Despite the low biodegradability of MF and UF-based shells, there is concern about the possible release of formaldehyde. Though it is not harmful once incorporated in the MC wall material, free formaldehyde is carcinogenic [40]. Therefore, it would be desirable to reduce or remove formaldehyde from aminoplast delivery systems.

One strategy could be the synthesis of MC with low formaldehyde-melamine molar ratio (F/M). Long et al. reported the preparation of microcapsules by using a low F/M, 0.2 to 0.49, in comparison with typical literature reported values, which are in the range of 2.3-5.5. According to the authors, it is possible to achieve a significant reduction in the levels of formaldehyde content using a F/M of 0.49, without affecting mechanical properties and encapsulation efficacy considerably [51].

As melamine can react with aldehydes to give aminoplast copolymers, another approach bases on the possibility of substituting formaldehyde by glyoxal and its derivatives. León et al. studied the mechanism of polymerization between glyoxal and its derivatives and melamine, urea or guanazole to prepare various oligomers of formaldehyde-free aminoplast resin. They prepared stable and robust capsules through a combined synthesis of an optimized composition of aminoplast copolymers followed by in situ cross-linking using hexamethylenediisocyanate [52].

In order to meet legal requirements, manufacturing companies add formaldehyde scavengers to lower and control free formaldehyde content in their products. That way, the risk of exposure to free formaldehyde is almost nonexistent [40]. However, on the one hand, the negative perception of consumers could prompt its replacement by bio-based polymers, due to their renewable feedstock and presumed biodegradability. It seems that not even different petroleum-based monomers could be accepted to build the MC walls. On the other hand, the use of bio-based polymers instead of the petroleum-based alternatives is not always acceptable on deeper sustainability reasons. In fact, a complete study of all factors with impact on the life-cycle analysis of a material would be required [40].

CONCLUSION

The versatility of melamine-formaldehyde resin as shell material for the encapsulation of diverse core materials was demonstrated through a broad spectrum of applications. A study of the use of MF-based MC loaded with linseed oil was conducted. The products obtained using four emulsifying agent mixtures to stabilize the emulsion and form robust MC were compared through optical microscopy and SEM. Conditioning operations were performed to isolate MC from the reaction medium with the aim to obtain a powder of MC, which could eventually be incorporated in the formulation of a protective coating. The most successful emulsifying combination for the production of LO-filled MF-based MC was the mixture of commercial unpurified GA in grain solution with SDS solution. The obtained MC were able to withstand the conditioning operations to yield a powder of MC. CRM studies allowed to verify effective encapsulation of core material. As regard the production of smaller MF-based capsules with the aid of an ultrasonic processor, further exploration of experimental conditions is required with the purpose of synthetizing a powder of MC. In addition, results about the synthesis of small UF-based MC with the aid of an ultrasonic homogenizer were included. Those MC were more resistant than MC prepared with the homogenizer and could endure preparation and conditioning procedures, allowing to obtain a powder of MC. Finally, despite the concern about possible emissions of formaldehyde from MF resins, it seems that they would continue being used responsibly until the mass arrival of more convenient materials from different points of view.

REFERENCES

[1] Kumar, Anil and Vimal Katiyar. 1990. "Modeling and Experimental Investigation of Melamine-Formaldehyde Polymerization." *Macromolecules* 23: 3729-3736.

[2] Meir Robert J., Andrew Tiller, and Sylvia A. M. Vanhommerig. 1995. "Molecular Modeling of Melamine-Formaldehyde Resins. 2. Vibrational Spectra of Methylolamines and Bridge Methylolamines." *J. Phys. Chem.* 99: 5457-5464.

[3] Yuan, Yan Chao, Min Zhi Rong, and Ming Qiu Zhang. 2008. "Preparation and Characterization of Microencapsulated Polythiol." *Polymer* 49: 2531-2541. doi: 10.1016/j.polymer.2008.03.044.

[4] Likozar, Blaž, Romana Cerc Korošec, Ida Poljanšec, Primož Ogorolec, and Peter Bukovec. 2011. "Curing Kinetics Study of Melamine-urea-formaldehyde Resin." *J. Therm Anal Calorim.* doi 10.1007/s10973-011-1883-0.

[5] Liu, Mingquan, Tri Minh Tran, Ahmed Awad Abbas Elhaj, Silje Bøen Torsetnes, Ole N. Jensen, Börge Sellergren, and Knut Irgum. (2017). "Molecurlaly Imprited Porous Monolithic Materials from Melamine-formaldehyde for Selective Trapping of Phosphopeptides." *Analytical chemistry* 89: 9491-9501. doi: 10.1021/acs.analchem.7b02470.

[6] Su, Jen-Feng, Sheng-Bao Wang, Jian-Wei Zhou, Zhen Huang, Yun-Hui Zhao, Xiao-Yan Yuan, Yun-yi Zhang, and Jin-Bao Kou. 2011. "Fabrication and Interphacial Morphologies of Methanol-melamine-formaldehyde (MMF) Shell MicroPCMs/epoxy Composites." *Colloid Polym Sci* 289: 169-177. doi: 10.1007/s00396-010-2334-3.

[7] Yin, Dezhong, Li Ma Wangchang Geng, Baoliang Zhang, and Qiuyu Zhang. 2015. "Microencapsulation of n-hexadecanol by In Situ Polymerization of Melamine-formaldehyde Resin in Emulsion Stabilized by Styrene-maleic Anhydride Copolymer." *International Journal of Energy Research* 39, 5: 661-667. First published December 29 2014. doi: 10.1002/er.3276.

[8] Alič B., U. Šebenik and M. Krajnc. 2012. "Microencapsulation of Butyl Stetearate with Melamine-formaldehyde Resin: Effect of Decreasing the pH Value on the Composition and Thermal Stability of Microcapsules." *Express Polymer Letters* 6: 826-836. doi: 10.3144/expresspolymlett.2012.88.

[9] Konuclu, Yeliz, Halime O. Paksoy, Murat Unal, and Suleyman Konuklu. 2014. "Microencapsulation of a Fatty Acid with Poly(melamine-urea-formaldehyde)." *Energy Conversion and Management* 80: 382-390. doi: 10.1016/j.enconman.2014.01.042.

[10] Zhang Huanzhi and Xiaodong Wang, 2009. "Fabrication and Performances of Microencapsulated Phase Change Materials based on n-octadecane Core and Resorcinol-modified Melamine-formaldehyde Shell." *Colloids and surface A: Physicochemical and Engineering Aspects* 332:129-138. doi: 10.1016/j.colsurfa.2008.09.013.

[11] Wang, Xianfeng and Tao Zhao. 2017. "Effects of Parameters of the Shell Formation Process on the Performance of Microencapsulated Phase Change Materials Based on Melamine-formaldehyde." *Textile Researc Journal* 87: 1848-1859. doi: 10.1177/0040517516659382.

[12] Wang Yan, Zhimin Liu, Xiaofeng Niu, and Xiang Ling. 2019. "Preparation, Characterization and Thermal Properties of Microencapsulated Phase Change Materials for Low-Temperature Thermal Energy Storage." *Energy Fuels* 33: 1631-1636. doi: 10.1021/acs.energyfuels.8b02504.

[13] Luo Wen-jun, Wei Yang, Shu Jiang, Jian-min Feng, and Ming-bo Yang. (2007). "Microencapsulation of Decabromodiphenil Ether by In Situ Polymerization: Preparation and Characterization." *Polymer Degradation and Stability* 92: 1359-1364. doi: 10.1016/j.polymdegradstab.2007.03.004.

[14] Wu, Ningjing, Zhaoxia Xiu, and Jiyu Du. 2017. "Preparation of Microencapsulated Aluminumm Hypophosphite and Flame Retardancy and Mechanical Properties of Flame-retardant ABS Composites." *Journal of Applied Polymer Science* 134: 45008-45021. First published May 7 2017. doi: 10.1002/app.45008.

[15] Du Xiaosheng, Yuanlai Fang, Xu Cheng, Zongliang Du, Mi Zhou, and Haibo Wang. 2018. "Fabrication and Characterization of flame-Retardant Nanoencapsulated n-octadecane with Melamine-formaldehyde Shell for Thermal Energy Storage." *ACS Sustainable Chem. Eng.* 6: 15541-15549. doi: 10.121/acssuschemeng.8b03980.

[16] Yuan Huizhu, Guangxing Li, Lijuan Yang, Xiaojing Yan, and Daibin Yang. "Development of Melamine-Formaldehyde Resin Microcapsules with Low Formaldehyde Emission Suited for Seed Treatment." *Colloids and surface B: Biointerfaces* 128: 149-154. doi: 10.1016/j.colsurfb.2015.02.029.

[17] Lee H. Y., S. J. Lee, I. W.Cheong, and J. H. Kim. 2002. "Microencapsulation of Fragant Oil via In Situ Polumerization: Effects of pH and Melamine-formaldehyde Molar Ratio." *J. Microencapsulation* 19, 5: 559-569. 10.1080/02652040210140472.

[18] Hwang, Jun-Seok, Jin-Nam Kim, Young-Jung Wee, Hong-Gi Jang, Sun-Ho Kim, and Hwa-Won Ryu. 2006. "Factors affecting the Characteristics of Melamine Resin Microcapsules Containing Fragant Oils." *Biotechnology and Bioprocess Engineering* 11: 391-395. doi: 10.1007/BF02932304.

[19] Sánchez Navarro, M. M., F. Payá Nohales, F. Arán Aís, and C. Orgilés Barceló. 2012. "Polymer Shell Nanocapsules Containng A Natural Antimicrobial Oil for Footwear Applications." *Prog Colloid Polym Sci* 139: 73-77. doi: 10.1007/978-3-642-28974-3_13.

[20] Fei, Xuening, Hongbin Zhao, Baolian Zhang, Lingyun Cao, Miaozhuo Yu, Jianguo Zhou, and Lu Yu. 2015. "Microencapsulation Mechanism and Size Control of Fragrance Microcapsules with Melamine Resin Shell." *Colloids and Surf. A: Physicochemical Engineering Aspects* 469: 300-306. doi: 10.1016/j.colsurfa.2015.01.033.

[21] Zhao, Hongbin, Xuening Fei, Lingyun Cao, Baolian Zhangc and Xin Liud. 2019. *RSC Adv.* 9: 25225–25231. Published August 13 2019. doi: 10.1039/c9ra05196a.

[22] Zhou, Y, Y. Yan, Y. Du, J. Chen, X. Hou, and J. Meng. "Preparation and Application of Melamine-formaldehyde Photocromic Microcapsules." *Sens. Actuators B Chem.* 188: 502-512. doi: 10.1016/j.snb.2013.07.049.

[23] Ming Meng, Ling, Yan Chao Yuan, Min Zhi Rong, and Min Qiu Zhang. 2010. "A Dual Mechanism Single-component Self-healing Strategy for Polymers." *J. Mater. Chem.* 20: 6030-6038. doi: 10.1039/c0jm00268b.

[24] Zhu, G., L. Lü, J. Tang, B. dong, N. Han, and F. Xing. 2013. "Preparation of Mono-sized Epoxy/MF Microcapsules in the Appearance of Polyvinyl Alcohol as Co-emulsifier." *ICSHM* 2013: Proceedings of the 4th International Conference on Self-Healing Materials. http://resolver.tudelft.nl/uuid:66947360-d3fb-4954-b710-aa22be96a19b.

[25] Sharma, Shilpi and Veena Choudhary. 2017. "Poly(melamine-formaldehyde) Microcapsules Filled with Epoxy Resin: Effect of M/F

Ratio on the Shell Wall Stability." *Mater. Res. Express* 4, 075307. doi: 10.1088/2053-1591/aa7c8f.

[26] Khorasani, Saied Nouri, Shahla Ataei, and Rasoul E. Neisiany. 2017. "Microencapsulation of a Coconut Oil-based Alkyd Resin into Poly(melamine-urea-formaldehyde) as Shell for Self-healing Purposes." *Prog. Org. Coat.* 111: 99-106. doi: 10.1016/j.porgcoat.2017.05.014.

[27] Suryanarayana, C., K. Chowdoji Rao, and Dhirendra Kumar. 2008. "Preparation and Characterization of Microcapsules Containing Linseed Oil and its Use in Self-healing Coating." *Prog. Org. Coat.* 63: 72-78. doi: 10.1016/j.porgcoat.2008.04.008.

[28] Selvakumar, N., K. Jeyasubramanian, and R. Sharmila. 2012. "Smart Coating for Corrosion Protection by Adopting Nano Particles." *Prog. Org. Coat.* 74: 461-469. doi: 10.1016/j.porgcoat.2012.01.011.

[29] Hasanzadeh, Majdeh, Mehdi Shahidi, and Maryam Kazemipour. 2015. "Application of EIS and EN Techniques to Investigate the Self-healing Ability of Coatings Based on Microcapsules Filled with Linseed Oil and CeO_2 Nanoparticles." *Prog. Org. Coat.* 80 (2015) 106-119. doi: 10.1016/j.porgcoat.2014.12.002.

[30] Hatami Boura S., M. Peikari, A. Ashrafi, and M. Samadzadeh. 2012. "Self-healing Ability and Adhesion Strength of Capsule Embedded Coatings-Micro and Nano Sized Capsules Containing Linseed Oil." *Prog. Org. Coat.* 75: 292-300. doi: 10.1016/j.porgcoat.2012.08.006.

[31] Nesterova, Tatyana, Kim Dam-Johansen, Lars T. Pedersen, and Søren Kiil. 2012. "Microcapsule-based Self-healing Anticorrosive Coatings: Capsule Size, Coating Formulation, and Exposure Testing." *Prog. Org. Coat.* 75, 4: 309-318. doi: 10.1016/j.porgcoat.2012.08.002.

[32] Lang, Sinuo and Qixin Zhou. 2017. "Synthesis and Characterization of Poly(urea-formaldehyde) Microcapsules Containing Linseed Oil for Self-healing Coating Development." *Prog. Org. Coat.* 105: 99-110. doi: 10.1016/j.porgcoat.2016.11.015.

[33] Wang, Haoran and Qixin Zhou. 2018. "Evaluation and Failure Analysis of Linseed Oil for Self-healing Anticorrosive Coating, *Prog. Org. Coat.* 118: 108-115. doi: 10.1016/j.porgcoat.2018.01.024.

[34] Behzadnasab, M., S. M. Mirabedini, M. Esfandeh, and R. R. Farnood. 2017. "Evaluation of Corrosion Performance of Linseed Oil-filled Microcapsules via Electrochemical Impedance Spectroscopy." *Prog. Org. Coat.* 105: 212-224. doi: 10.1016/j.porgcoat.2017.01.006.

[35] Khalaj Asadi, Amir, Morteza Ebrahimi, and Mohsen Mohseni. 2016. "Preparation and characterization of melamine-urea-formaldehyde microcapsules containing linseed oil in the presence of polyvinylpyrrolidone as emulsifier." *Pigment & Resin Technology* 46, 4: 318-326. doi: 10.1108/PRT-04-2016-0043.

[36] Abdipour Hamed, Mostafa Rezaei, and Farhang Abbasi. 2018. "Synthesis and Characterization of High Durable Linseed Oil-Urea Formaldehyde Micro/nanocapsules and their Self-healing behavior in epoxy coating." *Prog. Org. Coat.* 124: 200-212. doi: 10.1016/j.porgcoat.2018.08.019.

[37] Jadhav, Rajendra S., Vishal Mane, Avinash V. Bagle, Dilip G. Hundiwale, Pramod P. Mahulikar, and Gulzar Waghoo. 2013. "Synthesis of Multicore Phenol Formaldehyde Microcapsules and their Application in Polyurethane Paint Formulation for Self-healing Anticorrosive Coating." *Int. J. Ind. Chem.* 4: 31-39. First published May 11 2013. doi: 10.1186/2228-5547-4-31.

[38] Szabó Tamás, Judit Teledgi and Lajos Nyikos. 2015. "Linseed Oil-filled Microcapsules Containing Drier and Corrosion Inhibitor- Their Effects on Self-healing Capability of Paints." *Prog. Org. Coat.* 84: 136-142. doi: 10.1016/j.porgcoat.2015.02.020.

[39] Çömlekçi, Kurt G. and S. Ulutan. 2018. "Encapsulation of Linseed Oil and Linseed Oil Based Alkyd Resin by Urea Formaldehyde Shell for Self-healing Systems." *Prog. Org. Coat.* 121: 190-200. doi: 10.1016/j.porgcoat.2018.04.027.

[40] Bruyninckx Kevin and Michiel Dusselier. 2019. "Sustainable Chemistry Considerations for the Encapsulation Compounds in Laundry-Type Apllications." *ACS Sustainable Chem. Eng.* 7: 8041-8054. doi: 10.1021/acssuschemeng.9b00677.

[41] El Asbahani, A., K. Miladi, W. Badri, M. Sala, E. H. Aït Addi, H. Casabianca, A. El Mousadik, D. Hartmann, A. Jilale, F. N. R. Renaud,

and A. Elaissari. 2015. "Essential Oils: From Extraction to Encapsulation." *International Journal of Pharmaceutics* 483: 220–243. doi: 10.1016/j.ijpharm.2014.12.069.

[42] Radünz, Marjana, Elizabete Helbig, Caroline D. Borges, Tatiane K.V. Gandra, and Eliezer A. Gandra. 2018. A Mini-Review on Encapsulation of Essential Oils. *J Anal Pharm Res* 7, 1: 205-206. doi: 10.15406/japlr.2018.07.00205.

[43] Bakry, Amr M., Shabbar Abbas, Barkat Ali, Hamid Majeed, Mohamed Y. Abouelwafa, Ahmed Mousa, and Li Liang. 2016. "Microencapsulation of Oils: A Comprehensive Review of Benefits, Techniques, and Applications." *Comprehensive Reviews in Food Science and Food Safet.* 15: 143-182. doi: 10.1111/1541-4337.12179.

[44] Miguel M., R. Ollier, V. Alvarez, and C. Vallo. 2016. "Effect of the Preparation Method on the Structure of Linseed-oil-filled Poly (urea-formaldehyde) Microcapsules." *Prog. Org. Coat.* 97: 194-202. doi: 10.1016/j.porgcoat.2016.04.026.

[45] Miguel M. and C. Vallo. 2019. "Influence of the Emulsifying System to Obtain Linseed-oil-filled Microcapsules with a Robust Poly(melamine-formaldehyde)-based Shell." *Prog. Org. Coat.* 129: 236-246. doi: 10.1016/j.porgcoat.2019.01.026.

[46] Montenegro, Mariana. A., María L. Boiero, Lorena Valle, and Claudio D. Borsarelli. 2012. "Gum Arabic: More than an Edible Emulsifier." In *Products and Applications of Biopolymers*, edited by Reinhard Verbeek, chapter 1. IntechOpen, doi: 10.5772/33783. Available from: http://www.intechopen.com/books/products-and-applications-of-biopolymers/gum-arabic-more-than-an-edible-emulsifier.

[47] Sanchez, C., M. Nigen, V. Mejia Tamayo, T. Doco, P. Williams, C. Amine, and D. Renard. 2017. "Acacia gum: History of the future." *Food Hydrocolloids* 78: 140-160. doi: 10.1016/j.foodhyd.2017.04.008.

[48] Kong, H., J. Yang, Y. Zhang, Y. Fang, K. Nishinari, and G. O. Phillips. 2014. "Synthesis and antioxidant properties of gum-arabic stabilized selenium nanoparticles." *International Journal of Biological Macromolecules* 65: 155-162. doi: 10.1016/j.ijbiomac.2014.01.011.

[49] Tabatabaee Amid Bahareh and Hamed Mirhosseini. 2012. "Effect of different techniques on the characteristics of heteropolysaccharide-protein biopolymer from durian (Durio zibethinus) seed." *Molecules* 17, 9: 10875-10892. doi: 10.3390/molecules170910875.

[50] Larkin, Peter. 2011 Introduction to Infrared and Raman Spectroscopy. *Principles and Spectral Interpretation.* Chapter 1: Historical Perspective: IR and Raman Spectroscopy, edited by Peter Larkin, 1-5. Elsevier. http://www.sciencedirect.com/science/article/pii/B9780123869845100011.

[51] Long Yue, David York, Zhibing Zhang, and Jon A. Preece. 2009. "Microcapsules with low content of formaldehyde: preparation and characterization." *J. Mater. Chem.* 19: 6882-6887. doi: 10.1039/b902832c6882.

[52] León G., N. Paret, P. Fankhauser, D. Grenno, P. Erni, L. Ouali, and D. L. Berthier. 2017. "Formaldehyde-free Melamine Microcapsules as Core/Shell Delivery Systems for Encapsulation of Volatile Active Ingredients." *RSC Adv.* 7: 18962–18975. doi: 10.1039/c7ra01413a.

In: An Introduction to Melamine
Editor: Ashley Harris

ISBN: 978-1-53617-136-5
© 2020 Nova Science Publishers, Inc.

Chapter 4

MELAMINE SENSOR DEVELOPMENT BASED ON MIXED METAL OXIDE NANOPARTICLES

Mohammed M. Rahman[1,*], *Abdullah M. Asiri*[1] *and M. M. Alam*[2,#]

[1]Chemistry department, King Abdulaziz University,
Faculty of Science, Jeddah, Saudi Arabia
[2]Department of Chemical Engineering and Polymer Science,
Shahjalal University of Science and Technology (SUST),
Sylhet, Bangladesh

ABSTRACT

In this approach, a sensitive chemical sensor was developed to detect melamine selectively by electrochemical approach, where ternary mixed metal oxide ($ZnO/CuO/Co_3O_4$) nanoparticles (NPs) were prepared by wet-chemical process. The calcined $ZnO/CuO/Co_3O_4$ NPs were investigated by field emission scanning electron microscopy (FESEM), energy-dispersive

---

[*] Corresponding Author's Email: mmrahman@kau.edu.sa.
[#] Corresponding Author's Email: mmalamsust@gmail.com.

X-ray spectroscopy (EDS), X-ray photoelectron spectroscopy (XPS), powder X-ray diffraction (XRD), ultraviolet visible spectroscopy (UV-vis), and Fourier-transform infrared spectroscopy (FTIR). To the fabricate melamine sensor, slurry of NPs in ethanol was deposited as uniform thin layer on a glassy carbon electrode (GCE). The calibration curve of the proposed melamine sensor in form of current versus concentration (in logarithmic scale) plot is found to be linear over a melamine concentration range of 0.05 nM ~ 0.05 mM. The sensitivity of the sensor is very good and detection limit is very low. The melamine sensor with active $ZnO/CuO/Co_3O_4$ NPs shows good reliability, precise reproducibility and short response time in sensing performances. The developed ternary metal oxide nanoparticles based sensor is a new introduction in the sensor technology for determining melamine in environmental samples reliably.

LITERATURE REVIEW

Generally, melamine is a heterocyclic thiazine nitrogenous compound, which has received extreme attention in the sector of public health, due to the occurrences of renal failure and death in human; particularly in youngsters and animals [1, 2]. The major sources of human nutrition are milk and dairy products and they are of vital importance to growth and maintenance of human health. Considering the commercial aspect, the dairy products are illegally adulterated with melamine, urea, starch and glucose etc. [3-5]. Therefore, USA and EU implement a safety regulation that limits the melamine concentration in milk and dairy products to 2.5 and 1.0 ppm respectively [6-9]. Melamine is a nitrogenous compound and its adulteration in dairy foods enhances nitrogen contents [10, 11]. A number of researches have claimed that the ingestion of melamine via food chain is responsible kidney stones and subsequently renal failure in human [12]. Recent studies show that the toxicity of melamine is not only associated with renal injury, but also reproductive damage of skin, eye and respiratory system [13-16]. Beside this, melamine has been to disrupt blood-testis barrier in mice even at a low-dose level [17]. High melamine concentration can reduce DNA production rate in animals [18]. Therefore, control of melamine is very important to assure food safety.

A number of analytical methods are applied to detect melamine in aqueous media. Among the methods widely used are surface enhanced Raman spectroscopy [19], gas chromatography [20], enzyme colorimetric assay [21], liquid chromatography [22], and high performance liquid chromatography/mass spectrometry [23]. These techniques however, are not always appealing due to many disadvantages. Among them notables are heavy and complicated instrumentation, high cost, high time consumption and hard transfermability [24]. Therefore, there is still high demand for melamine detection technique with high sensitivity, faster response, simple instrumentation, convenience and reliability. To overcome the drawback of the mentioned analytical methods, the electrochemical technique with good sensitivity, simplicity, long-term stability and broad range of detection has been investigated to analysis melamine in aqueous medium [25, 26]. For electrochemical sensor, doped or non-doped metal oxides are used as electron mediator for sensing melamine in aqueous medium. A number of metal oxides such as Au-Fe_3O_4 [27], Cd-doped Sb_2O_4 [28], silver NPs [29], Pt-ZnO [30], and ZnO NPs [31] has been investigated for the detection of melamine. With 3.37 eV band gap and 60.0 meV excitation binding energy, the p-type semi-conductor ZnO, has attracted high attention for its stability, large specific surface area, photosensitivity and intrinsic electronic properties to be used as sensing element [32-34] and it has been studied as effective sensing elements for glycine [35], benzaldehyde [36] xanthine [37], 4-aminophenol [38], ethanol [39], acetone [40] and so on. CuO is also a p-type semiconducting metal oxide, and due to its excellent magnetic, electronic and optical properties, it has got potential application as catalyst, sensor, semiconductor, supercapacitor, magnetic storage and infrared filters [41]. As electrochemical sensor, CuO has been found to detect 1,2-dichlorobenzen [42], Kanamycin [43] and 2-nitrophenol [44] in phosphate buffer medium. Magnetic Co_3O_4 with its appealing physical and electronic properties, has also proved to be an efficient candidate using as sensor element [45, 46]. Thus, it is quite obvious that each of the three semi-conductive metal oxides, namely; ZnO, CuO and Co_3O_4, individually has been successfully tested as sensing elements in various electrochemical sensors.

In this study, we have developed an electrochemical sensor based on ternary combination of ZnO/CuO/Co$_3$O$_4$ NPs and GCE. Wet-chemically prepared ZnO/CuO/Co$_3$O$_4$ NPs have been coated on GCE with conductive 5% nafion binder and the resulted assembly is implemented as melamine sensor. The prepared melamine chemical sensor has exhibited good sensitivity, a broad linear dynamic range, lower detection limit, reliable reproducibility, short response time and consistently stability in aqueous medium. It is concluded that ZnO/CuO/Co$_3$O$_4$ nanomaterial is promising candidate for the development of electrochemical sensor.

EXPERIMENTAL APPROACH

Materials and Methods

Analytical grade chemicals reagents Co(NO$_3$)$_2$•6H$_2$O, Zn(NO$_3$)$_2$•6H$_2$O and CuCl$_2$•2H$_2$O collected from Sigma-Aldrich Co. (USA) were used to prepare nanomaterial. To accomplish this study, a number of environmental toxics including zimtaldehyde, 2,4-dinitrophenol, melamine, 2-acetylpyridine, benzylchloride, diethylmalonate, methylcyanite, M-xylol, and M-tolyhydrazine were collected from Sigma-Andrich. Monosodium disodium and disodium phosphate buffer and nafion suspension in ethanol were prepared from chemicals available in the department store. The optical property of the calcined ZnO/CuO/Co$_3$O$_4$ NPs was studied by UV-vis. Spectroscopy (300 UV-vis. Spectrophotometer, Thermo-Scientific) and FTIR spectrometer (thermos-scientific NICOLET iS50, Madison, WI, USA). The binding energies of Zn, Cu, Co and O were evaluated by XPS analysis on a K-α1 spectrometer (Thermo scientific, K-α1 1066) with an excitation radiation source (A1 Kα1, Beam spot size = 300.0 μm, pass energy = 200.0 eV, pressure~ 10-8 torr). The structural and elemental analysis of the synthesized NPs was done by FESEM (JEOL, JSM-7600F, Japan) equipped with EDS. The crystallinity of the NPs was determined by powder X-ray diffraction (XRD) with the ARL™ X'TRA powder diffractometer. The electrochemical (I-V) investigation was carried out by Keithley electrometer (6517A, USA).

Synthesis of ZnO/CuO/Co$_3$O$_4$ NPs by Wet-chemical Process

The chemicals, Co(NO$_3$)$_2$•6H$_2$O, Zn(NO$_3$)$_2$•6H$_2$O and CuCl$_2$•2H$_2$O were used to prepare 0.1 M solution of each in deionized water in three separate volumetric flasks of 100.0 mL. To synthesized nanomaterials, 50.0 mL from each of the prepared solutions was poured in a 250.0 mL conical flask. Then, the flask was placed on a hot plat at 80.0°C. Subsequently, a 0.1 M NH$_4$OH was added, with continuous magnetic stirring, to the solution in conical flask dropwise the pH of the solution reached 10.5. At this condition, all the metal ions co-precipitated out quantitatively in form of metal hydroxide. Then, the system is kept at standstill for several hours for the co-precipitates to settle down.

The reactions scheme in the ternary mixture is supposed to be that as described below (Eqs. (i-v)).

$$NH_4OH_{(l)} \leftrightarrows NH_4^+{}_{(aq)} + OH^-{}_{(aq)} \tag{i}$$

$$Zn(NO_3)_{2(s)} \rightarrow Zn^{2+}{}_{(aq)} + 2NO_3^-{}_{(aq)} \tag{ii}$$

$$CuCl_{2\,(s)} \rightarrow Cu^{2+}{}_{(aq)} + 2Cl^-{}_{(aq)} \tag{iii}$$

$$Co(NO_3)_{2(s)} \rightarrow Co^{2+}{}_{(aq)} + 2NO_3^-{}_{(aq)} \tag{iv}$$

$$Zn^{2+}{}_{(aq)} + Cu^{2+}{}_{(aq)} + Co^{2+}{}_{(aq)} + OH^-{}_{(aq)} + nH_2O$$
$$\leftrightarrows Zn(OH)_2 \bullet Cu(OH)_2 \bullet Co(OH)_{2\,(s)} \bullet nH_2O \downarrow \tag{v}$$

The precipitation of the metal ions in wet-chemical method depends on the value of solubility product constant (K_s) of the respective metal hydroxide. According to the hand book of physical chemistry, Ks-value of Cu(OH)$_2$, Co(OH)$_2$ and Zn(OH)$_2$ are 2.2*10^{-20}, 5.92*10^{-15} and 3.0*10^{-17}

respectively [47]. With the continuous dropwise addition of NH_4OH, the concentration of OH^- in the solution increases. Therefore, $Cu(OH)_2$ with lower K_s-value among the three starts precipitating and forming nuclei of crystal. Then, the crystallites of $Cu(OH)_2$ aggregates with one another forming larger crystallites. As the pH of solution increases continuously, $Zn(OH)_2$ with the 2nd lower value of K_s, precipitates and is adsorbed on the crystallites of $Cu(OH)_2$. At higher pH, the 3rd metal Co^{2+} with the lower K_s-value of its hydroxide co-precipitates out in the same way. Similar nanocrystal formation has been reported by previous authors [39, 40, 46]. The resultant crystals of nanomaterials are filtered out from water and subsequently, washed with water and acetone. After that, the washed nanocrystals are kept inside an oven at 110.0°C temperature for overnight. Then, the dried nanocrystals are subjected to calcination in a high temperature muffle furnace at 500°C for 6 hours. During the calcination, the metal hydroxides are transformed into the corresponding oxides (Eq. (vi)).

In the muffle furnace:

$$Zn(OH)_2 \bullet Cu(OH)_2 \bullet Co(OH)_{2\,(s)} + O_2 \rightarrow ZnO \bullet CuO \bullet Co_3O_4 + H_2O_{\,(v)} \quad (vi)$$

Fabrication of GCE Working Electrode

The electrochemical sensor is a tiny device based on nanomaterial of metal oxide or composites that performs the function of a working electrode. To prepare the working electrode, a glassy carbon electrode (GCE) with surface area of 0.0316 cm^2 was coated with slurry of ZnO/CuO/Co$_3$O$_4$ NPs in form of a thin uniform film and was allowed to dry at ambient condition. To ensure tong-term stability of the electrode, a drop of nafion suspension (5% nafion in ethanol) known as conducting binder was added on film. Then, the electrode was placed inside an oven at 35.0°C for 24 hours, a pre-determined period of time that was found to be long enough for complete evaporation of ethanol leaving the electrode completely dry. An electrical circuit was assembled with Keithley electrometer, where ZnO/CuO/Co$_3$O$_4$ NPs/binder/GEC performed as a working and a Pt-wire as counter electrode.

A melamine solution with concentration of 0.5 mM was used as stock solution, from which a number solution in concentration range of 0.05 nM~0.05 mM was prepared by dilution. To estimate the analytical performances of the sensor, a calibration curve as current vs. concentration of melamine was plotted. The linear dynamic range (LDR) was determined from the maximum linearity (highest value of R^2, regression co-efficient) of the I vs. $\ln C$ plot (where I is the current and C is the melamine concentration). The sensor sensitivity was calculated from the slope of calibration curve divided by the surface area of the GCE. The detection limit (DL) of the sensor was computed at the signal to noise ratio of 3.

RESULT AND DISCUSSION

Characterization of ZnO/CuO/Co$_3$O$_4$ NPs

The crystal structure of the synthesized NPs of ZnO/CuO/Co$_3$O$_4$ was evaluated by powder X-ray diffraction. The morphology and the optical properties were investigated by the UV-vis, FTIR, FESEM, TEM, and EDS analysis. The bonding energy (eV) of individual elements existing in prepared NPs was quantified by XPS analysis.

Morphological and Elemental Analysis

The morphological (structural) investigation of aggregated ZnO/CuO/Co$_3$O$_4$ NPs was performed by FESEM. The resulted magnified image is represented in Figure 1(a to c). As it is observed from the Figure 1(c) (inset: high magnified image), the synthesized metal oxides are in spherical shapes with nano-level diameter. Similar picture is observed in EDE analysis as well (Figure 1d). The composition (Figure 1e) of the prepared ZnO/CuO/Co$_3$O$_4$ NPs is 54.25% O, 22.01% Co, 14.82% Cu, and 8.92% Zn. The peaks associated with any impurity are not visible in the elemental analysis of the EDS. Therefore, the synthesized ZnO/CuO/Co$_3$O$_4$ NPs consist of O, Cu, Co, and Zn only.

In additional EDS mapping study, the elemental dispersion was investigated and presented in the Figure 2. The analysis of the elemental composition was executed with respect to several different locations as shown in SEM images (Figure 2a). For ZnO/CuO/Co$_3$O$_4$ NPs (Figure 2a), Oxygen (Figure 2b), Zn (Figure 2c), Cu(Figure 2d), and Co(Figure 2e), the corresponding results are presented separately in each elemental images of Figure 2. From the elemental mapping investigation, it is clearly seen that elements are properly distributed in the aggregated ZnO/CuO/Co$_3$O$_4$ nanoparticle materials.

Figure 1. Morphological and elemental analysis of ZnO/CuO/Co$_3$O$_4$ NPs, (a-b) low and high magnifying images of FESEM investigation (c-inset: spherical nanoparticle), and (d-e) elemental analysis of ZnO/CuO/Co$_3$O$_4$ NPs and their corresponding spectrum with elemental ratios.

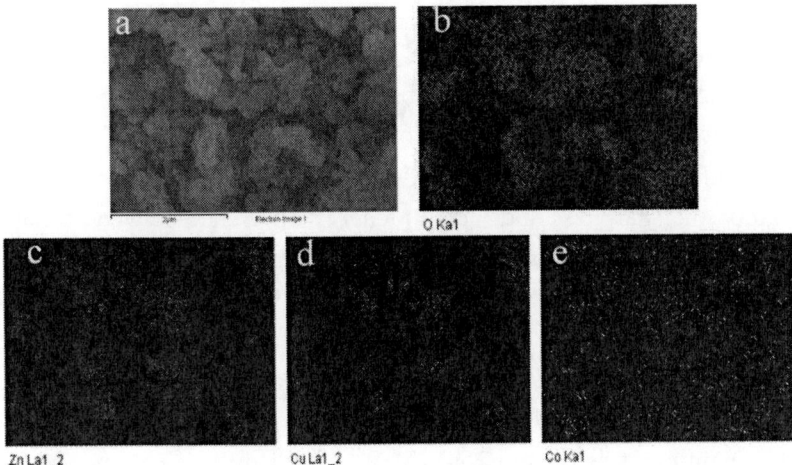

Figure 2. (a-e). Elemental mapping analysis of aggregated ZnO/CuO/Co$_3$O$_4$ NPs.

XPS Analysis of ZnO/CuO/Co$_3$O$_4$ NPs

The oxidation states and the corresponding binding energy of the compositional elements in ZnO/CuO/Co$_3$O$_4$ NPs are shown in Figure 3, and this investigation was executed by the implementations of XPS analysis. As it is presented in Figure 3(a), the XPS spectra of Zn2p shows two peaks at 1020.75 eV for Zn2p$_{3/2}$ and 1042.5 eV for Zn2p$_{1/2}$ with spin energy separation of 21.75 eV. Both the peaks of Zn2p orbit represent equal oxidation state of Zn^{2+} as reported by previous authors [48-50]. The XPS spectra of O1s as shown in Figure 3(b), is centered at 531.50 eV and the binding energy is related with O^{2-} ion at oxygen deficient region in ZnO [51-53]. As it is perceived from the Figure 3(c), the core level Cu2p XPS spectra shows two peaks at 934.5 eV and 954.5 eV corresponding to Cu2p$_{3/2}$ and Cu2p$_{1/2}$ respectively with spin-orbit splitting of 20.0 eV. Beside this, the two satellite peaks of Cu2p3/2 and Cu2p1/2 are centered at 938.0 eV and 957.0 eV respectively (spin energy separation is 19.0 eV). Therefore, both the peaks (core level and satellite) confirm the existence of Cu^{2+} oxidation state [54-57]. Beside this, Co2p spectra has two main peaks with the binding energy of 781.0 eV for Co2p$_{3/2}$ and 796.0 for Co2p$_{1/2}$ respectively and there are indication of the presence of Co^{3+}. The spin energy difference between the two peaks is 15.0 eV; which is in good agreement with the observation of the previous authors [58]. As it is observed in Figure 3(d), the XPS spectra of Co2p shows two satellite peaks along with the main peaks. Therefore, it is really difficult to ascertain the oxidation state of Co2p. To diffuse the ambiguity, the sharpness and the spin energy difference between satellites and main peaks of Co2p could be taken into consideration for identifying the oxidation state of Co. As it is seen in Figure 3(d), two satellite peaks are centered at 786.5 eV and 803.0 eV corresponding to $Co2p_{3/2}$ and $Co2p_{1/2}$ spin orbits respectively, and this value of satellite peaks could be ascribed to Co^{2+} [59] and it is detected that the spin energy difference between the satellite and the main peaks is around 15.0 eV, it presumed that Co^{2+} and Co^{3+} co-exist in the synthesized $ZnO/CuO/Co_3O_4$ NPs [60-62].

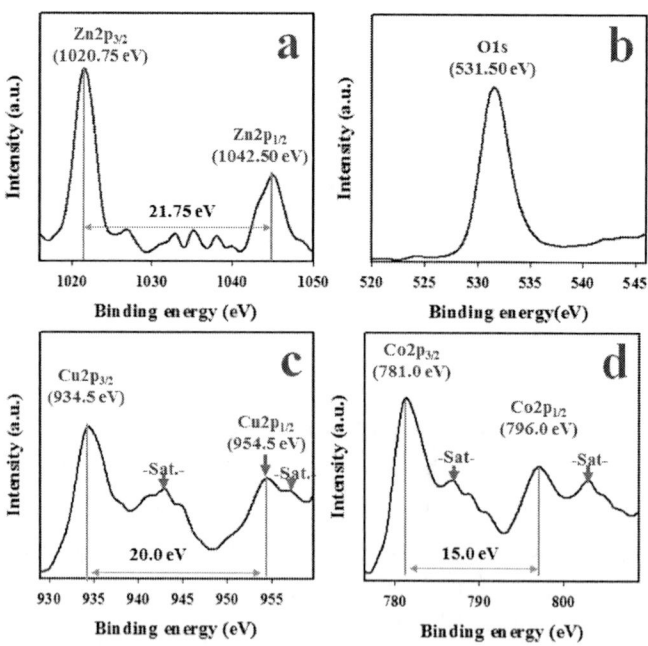

Figure 3. Surface composition and oxidation states of $ZnO/CuO/Co_3O_4$ NPs, (a) the core level high resolution Zn2p spin orbit, (b) O1s orbit, (c) spin orbit Cu2p level, and (d) Co2p spin orbit of $ZnO/CuO/Co_3O_4$ NPs.

Optical and Structural Characterization of ZnO/CuO/Co₃O₄ NPs

The phase crystallinity of the NPs was investigated by the powder X-ray diffraction with Kα radiation ($\lambda = 0.15418$ nm) source, and the peaks were identified at 2θ. The resultant XDR pattern presented in Figure 4(a) shows crystalline phases of ZnO, CuO and Co_3O_4 only. A number of crystalline planes of ZnO (marked as θ) are identified as (100), (002), (101), (102), (110), (103), and (201). The marked planes of ZnO have got much similarities with JCPDS no. 05-0664 and those reported by previous authors [63-66]. From the XRD diffraction spectrum in Figure 3(a), the phases of CuO (marked as β) are identified as (110), (002), (111), (202), (020), (202) and (113) planes. These peaks of CuO are in agreement with those reported in articles [67-69] and as well as in JCPDS no. 045-0937. The phases of Co_3O_4 are also noticed in XRD pattern in Figure 3(a) and are marked as a. The crystalline planes of Co_3O_4 are clearly identified as (220), (311), (422), (511), and (440); which are in good agreement with those in JCPDS no. 042-14687. Previous few authors have also noticed similar planes of the crystalline Co_3O_4 phases [70, 71].

The particles size of the NPs is estimated by the Scherer equation (Eq. (vii) as follows:

$$D = 0.9\lambda/(\beta \cos\theta) \tag{vii}$$

where, λ is wavelength (X-ray radiation = 1.5418 Å), β is the half of the width of highest peak, and θ is the diffracted angle. Applying the Eq. (vii) on the diffracted data at (101) plane of ZnO, the particle size is calculated to be 21.80 *nm*.

The UV-vis spectroscopy method was applied on the synthesized ZnO/CuO/Co₃O₄ NPs and the band gap energy (eV) has been estimated. The UV-vis spectrum is a result of the absorption of visible light source by nanomaterial and the transfer of the electrons from lower to higher energy level. Such UV-vis spectrum of ZnO/CuO/Co₃O₄ NPs is represented in Figure 4(b). As in the Figure, a broad and intense absorption peak appears in the neighborhood of 290.0 nm, which is characteristic for ZnO/CuO/Co$_3$O$_4$ NPs. This absorption band has appeared due to the transition of valence band electron of ZnO/CuO/Co$_3$O$_4$ NPs. The band gap energy E_{bg} is calculated by the Eq. (viii) and it is found to be 4.28 eV [72-74].

$$E_{bg} = 1240/\lambda_{max} \tag{viii}$$

where, E_{bg} is band gap energy, λ_{max} is maximum absorption wave length.

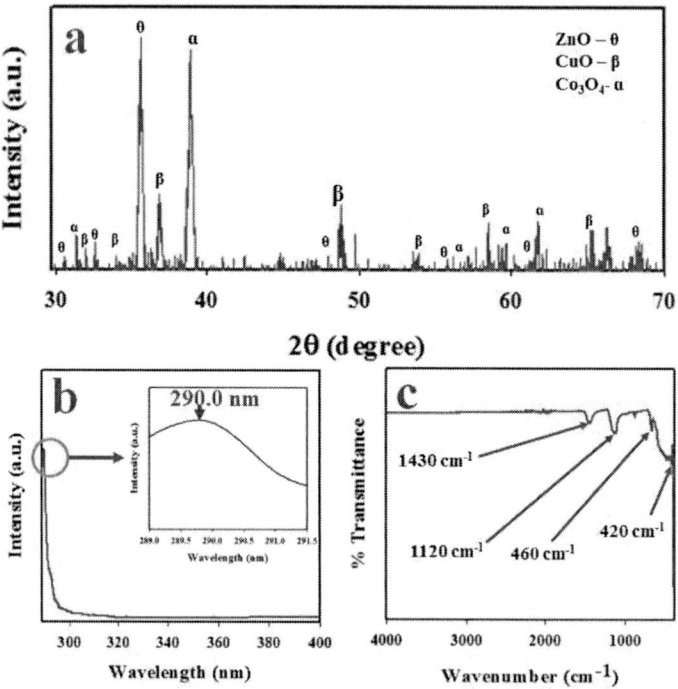

Figure 4. Optical behavior, crystallinity and phase evaluation of ZnO/CuO/Co$_3$O$_4$ NPs. (a) X-ray diffraction of ZnO/CuO/Co$_3$O$_4$ NPs, (b) UV-vis spectrum, and (c) FTIR spectrum.

The FTIR investigation was applied on ZnO/CuO/Co$_3$O$_4$ NPs in range of 400-4000 cm^{-1} as illustrated in Figure 4(c). The resulted FTIR spectrum shows four identical infrared absorption band at 420, 660, 1120, and 1430 cm^{-1}, and the obtained peaks at 420 and 660 cm^{-1} correspond to the stretching mode of Zn-O and Cu-O respectively [75-78]. The observed absorption band at 1120 and 1430 cm^{-1} are ascribed to the stretching modes of C-O and C=O respectively [89-81].

Optimization of Sensor Performances

The electrochemical sensor prepared in this work consists of ZnO/CuO/Co$_3$O$_4$ NPs coated on GCE, in which the NPs were bound to GCE tightly by the addition of Nafion. The nafion is a co-polymer and commercially available as 5% suspension in ethanol. The used nafion not only improves the binding strength, but also increases the electron transfer rate and conductance of the working electrode performing the function of chemical sensor [82, 83]. To ensure the maximum efficiency of the chemical sensor based on ZnO/CuO/Co$_3$O$_4$ NPs/binder/GCE, the pH dependence of the I-V response was studied first. As in the Figure 5(a), the highest I-V response of the proposed chemical is obtained at pH 5.7. Then, a number of environmental toxic chemicals with concentration of 5.0 nM were analyzed by the ZnO/CuO/Co$_3$O$_4$ NPs/binder/GCE chemical sensor at pH 5.7. The electrochemical responses (I-V) of zimtaldehyde, 2,4-dinitrophenol, melamine, 2-acetylpyridine, benzylchloride, diethylmalonate, methylcyanite, M-xylol, and M-tolyhydrazine are represented in Figure 5(b). Obviously, melamine exhibits the highest I-V response among all the toxic chemicals. On the basis of highest I-V response, melamine is considered to be the selective toxic chemical for the assembled electrochemical sensor with ZnO/CuO/Co$_3$O$_4$ NPs/binder/GCE. Afterwards, a series of melamine solutions in the concentration-range of 0.5 mM to 0.05 nM were analyzed and the results are presented in Figure 5(c). This performance was executed at the applied potential of 0~+1.5 V. Obviously, the electrochemical response increases from the lower to the higher concentration of melamine, and the response curves are completely distinguishable from one another. Similar experiences have been reported in detecting toxic chemicals by electrochemical approaches (I-V method) in our previous works [38, 84]. To evaluate the analytical performances such as sensitivity, linear dynamic range (LDR) and detection limit (DL), the resultant current data from Figure 5(c) at applied potential +1.5 V are isolated and plotted as current vs. conc. of melamine in Figure 4(d) known as calibration curve for the projected melamine sensor. The sensitivity of melamine chemical sensor is calculated from the slope of calibration curve and surface area of GCE (0.0316 cm^2) and it is found to be 36.98 µAµM^{-1}cm^{-2}. Obviously, this is an appreciable outstanding sensitivity. As it is observed from the inset in Figure 5(d), the current vs. logarithm of concentration plot is linear (regression co-efficient R^2=0.9958) in the range of 0.05 nM to 0.05 mM, and this concentration range is designed as linear dynamic range (LDR). Definitely, the LDR of melamine for prepared sensor is a highly wide range of concentration. The detection limit (DL) is computed from the calibration curve for the signal to noise ratio of 3, and the estimated value is found to be 9.7 ± 0.5 pM, a concentration-value low enough to be considered highly satisfactory. Taking into consideration the high sensitivity (36.98 µAµM^{-1}cm^{-2}) of the projected melamine chemical sensor, it may be predicted that the melamine chemical sensor based on ZnO/CuO/Co$_3$O$_4$ NPs/binder/GCE will show high catalytic decomposition ability and adsorption capacity as discussed in previous reports [85-87]. Thus, it is reasonable to except that ZnO/CuO/Co$_3$O$_4$ NPs with high crystallinity and the particle size as small as 21.80 nm, when used as chemical sensor would provide appreciable nano-atmosphere for reliable detection of melamine in phosphate buffer.

Response time is also an important parameter for the evaluation of analytical performances of an electrochemical sensor, and it is defined as the time required for the current vs. time curve for a given applied voltage to reach 95% of the ultimate current-value. The response time was measured at +1.5 V for 5.0 nM melamine solution in phosphate buffer and the corresponding result is illustrated in Figure 6(a). As seen in the Figure, steady state response of melamine is achieved in 15.0 sec by the projected melamine chemical sensor, and with some reserve for variation in melamine concentration and indeterminate factors, the response recording time could be chosen to be 17.0 sec, and all the current data described in this work have been recorded at least 17s after the measurement circuit is closed.

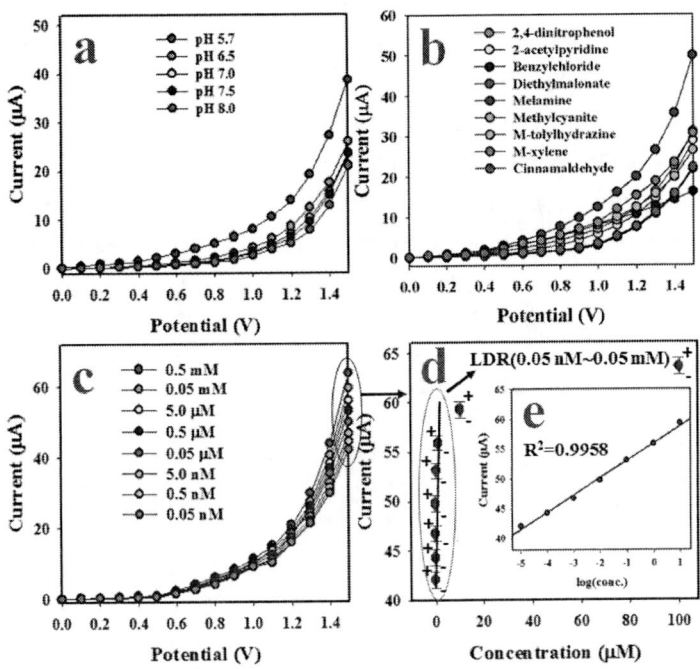

Figure 5. Optimization of sensor analytical performances of ZnO/CuO/Co$_3$O$_4$ NPs/binder/GCE electrode, (a) pH optimization, (b) selectivity test with analyte concentration 5.0 nM, (c) I-V responses to melamine concentration, and (d) exploration of calibration curve [inset: log(conc.) vs current].

The projected melamine sensor was subjected to interference test with some other chemicals. As seen in Figure 6(b), the I-V responses of melamine sensor do not deviate appreciably from its original value in presence of urea and glucose, individually or combined. Therefore, it can said that the projected melamine sensor does not face any interference effect from other chemicals. To evaluate the reproducibility of the I-V responses of the melamine sensor, the experiment was repeated seven times with melamine concentration of 5.0 nM and applied potential 0~+1.5V in phosphate buffer medium. As seen in Figure 6(c), the data from the seven replicated runs are practically indistinguishable from one another. The electrochemical responses do not change appreciably even for washing of the electrode after each run. Therefore, this reproducibility test provides evidence for reliability of the method. To estimate the precision of reproducibility test, the relative standard deviation of I-V runs are calculated at applied potential +1.5V, and it is found to be 0.88%. The reproducibility of the performance test is high enough to perceive that the projected melamine chemical sensor is efficient to detect melamine in real field of application. A control experiment has been performed separately by $ZnO/CuO/Co_3O_4$ NPs/binder/GCE sensor probe with and without glucose, melamine, starch, urea. It is observed the highest current response towards the melamine with $ZnO/CuO/Co3O4$ nanoparticles sensor probe compared to other analytes.

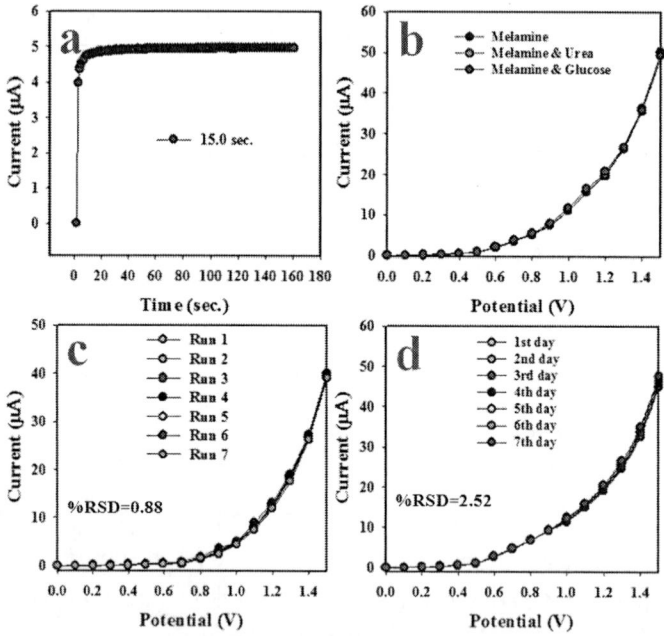

Figure 6. Evaluation of reliability performance of sensor $ZnO/CuO/Co_3O_4$ NPs/binder/GCE, (a) Response time for melamine concentration 5.0 nM, (b) interference effect test (Concentration: 5.0 nM melamine, 5.0 nM urea, 5.0 nM glucose), (c) reproducibility test with melamine concentration 5.0 nM, and (d) long time performance test.

The reproducibility test was executed for longevity (seven days) as well and the results are presented in Figure 6(d). Similar findings are observed in seven replicate runs in a single day (Figure 6(c)). Therefore, it is quite expected that the projected chemical sensor is able to perform long-period with similar efficiency.

As shown in Figure 5(c), at a given potential, the observed current is proportional to the concentration of melamine (in logarithmic scale). Therefore, an increase in I-V response is observed with increasing of analyte concentration. Such trend of electrochemical response has been reported by previous authors [88-90]. During the electrochemical analysis, melamine is oxidized through an oxidative passageway. In this process, the amino groups of melamine are oxidized to –NHOH and subsequently transformed into –OH group with the release of NH_3. Thus, melamine is converted to ammeline first, and with progress of the oxidation reaction, ammelide is produced from ammeline. Finally, cyanuric acid is produced in slightly acidic medium of pH 5.6. Similar oxidation mechanism of melamine has been described in earlier reports [91, 92]. The oxidation mechanism of melamine is illustrated in Scheme 1. As observed from Scheme 1, the electrochemical oxidation of melamine releases electron increasing the conductivity of the sensing medium.

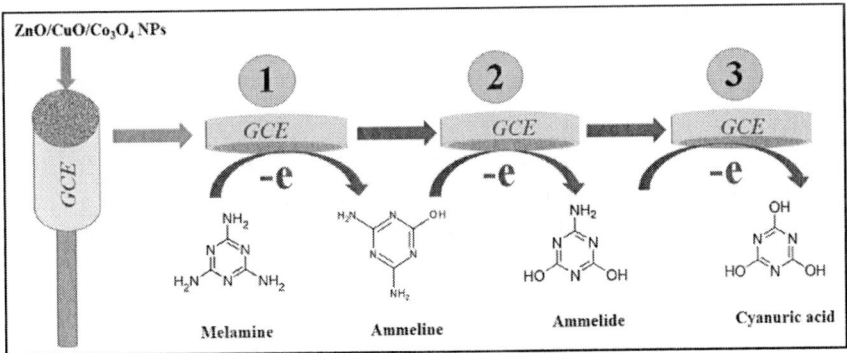

Scheme 1. Proposed sensing mechanism to detect melamine by $ZnO/CuO/Co_3O_4$ NPs/binder/GCE.

At the very beginning of the sensing process, a few number of melamine molecules are adsorbed on the surface of the thin film of ZnO/CuO/Co$_3$O$_4$ NPs/binder/GCE and the oxidation reaction is initiated. In this condition, the surface coverage on ZnO/CuO/Co$_3$O$_4$ NPs/binder/GCE is inadequate. With the progress of the adsorption process, the concentration on the electrode surface increases (increasing the surface coverage), and consequently, the oxidation rate increases. Finally, the surface coverage attains a saturation value (which is in equilibrium with the melamine concentration present in the solution), and at this state, the highest steady state current for the given concentration is obtained. As observed in Figure 6(a), the system reaches the equilibrium state in 15s. Also, it was found that the projected chemical sensor does not face any interference effect from other chemicals in the sensing medium (shown earlier in Figure 6(b)). Therefore, the projected chemical sensor is selective to melamine only. A comparison between the sensor performances of the projected one and a similar one reported in literature [28] has been presented in Table 1. As seen in the Table 1, the projected melamine chemical sensor based on ZnO/CuO/Co$_3$O$_4$ NPs/binder/GCE is superior to the relevant one based on Cd-doped Sb$_2$O$_4$ NPs/GCE in terms of all the three most important performance parameters (sensitivity, linear dynamic range and detection limit).

Table 1. Sensor performance of the projected one as compared to that of a similar one reported in literature

Modified GCE	DL	LDR	Sensitivity	Ref.
Cd-doped Sb$_2$O$_4$ NPs/GCE	14.0 pM	0.05 nM–0.5 mM	3.153 µAµM^{-1}cm^{-2}	[28]
ZnO/CuO/Co$_3$O$_4$ NPs/GCE	9.7 pM	0.05 nM-0.05 mM	36.98 µAµM^{-1}cm^{-2}	This works

* DL (detection limit) and LDR (linear dynamic range).

Recovery Test of the Sensor with Real Samples

Real environmental samples were investigated by the projected melamine sensor applying recovery method. Following this method, a known concentration of melamine solution was added to real environmental sample. The real samples were collected from various sources such as extract from PC-baby bottle, PVC-food packaging bag, PVC-water bottle and waste effluent from industry. The analysis results are represented in Table 2 and appear to be quite satisfactory.

Table 2. Analyses of real environmental samples with ZnO/CuO/Co$_3$O$_4$ NPs/binder/GCE sensor

Real Samples	Added melamine conc. (nM)	Measured melamine conc.[a] by ZnO/CuO/Co$_3$O$_4$ NPs/GCE (nM)			Average recovery[b] (%)	RSD[c] (%) (n = 3)
		R1	R2	R3		
Industrial effluent	5.000	4.725	4.720	4.856	95.34	1.63
PC- baby bottle	5.000	4.949	4.936	4.963	98.99	0.28
PVC- water bottle	5.000	4.983	4.961	5.009	99.68	0.49
PVC- food packaging bag	5.000	4.975	4.981	4.967	99.48	0.14

[a]Mean of three repeated determination (signal to noise ratio 3) with ZnO/CuO/Co$_3$O$_4$ NPs /GCE.
[b]Concentration of melamine determined/Concentration taken. (Unit: nM).
[c]Relative standard deviation value indicates precision among three repeated measurements (R1,R2,R3).

CONCLUSION

In this approach, glassy carbon electrode (GCE) has been successfully coated by a thin film of ZnO/CuO/Co$_3$O$_4$ NPs with the adhesive support of nafion, and is tested as a chemical sensor. The sensor has short response time, reliable reproducibility, and long-term stability, and is selective to melamine in phosphate buffer. The current vs. concentration (in logarithmic scale) relation for the melamine sensor at a given applied voltage is found to be linear in a wide range. The determination of melamine with this sensor is free of interference from other chemicals as urea and glucose. The proposed melamine sensor exhibits good sensitivity, broad linear dynamic range, and low detection limit. The ZnO/CuO/Co$_3$O$_4$ NPs-based GCE is a new introduction in the sensor technology and can be reliably employed in detecting melamine in real samples.

REFERENCES

[1] Khalil, S. R., A. Awad and S. A. Ali, *Environ. Toxicol. Pharmacol.*, 2017, 50, 136–144.
[2] Liu, J. M., A. Ren, L. Yang, J. Gao, L. Pei, R. Ye, Q. Qu and X. Zheng, *CMAJ,* 2010, 182, 439–443.
[3] Ma, Y., W. Dong, H. Bao, Y. Fang and C. Fan, *Food Chem.,* 2017, 221, 898–906.
[4] Jawaid, S., F. N. Talpur, S. T. H. Sherazi, S. M. Nizamani and A. A. Khaskheli, *Food Chem.,* 2013, 141, 3066–3071.
[5] Wu, Q., Q. Long, H. Li, Y. Zhang and S. Yao, *Talanta*, 2015, 136, 47–53.
[6] Ai, K. L., Y. L. Liu and L. H. Lu, *J. Am. Chem. Soc.,* 2009, 131, 9496–9496.
[7] Gu, C., T. Lan, H. Shi and Y. Lu, *Anal. Chem.*, 2015, 87, 7676–7682.
[8] Xu S. and H. Lu, *Biosens. Bioelectron,* 2015, 73, 160–166.
[9] Dai, H. C., Y. Shi, Y. L. Wang, Y. J. Sun, J. T. Hu, P. J. Ni and Z. Li, *Sens. Actuators B,* 2014, 202, 201–208.
[10] Shan-Shan, L., Y. Xin-Yao and W. Jian-Xiu, *Chin. J. Anal. Chem.,* 2014, 42, 695–700.
[11] Tyan, Y. C., M. H. Yang, S. B. Jong, C. K. Wang and J. Shiea, *J. Anal. Bioanal. Chem.,* 2009, 395, 729–735.
[12] Yin, R. H., X. T. Li, X. Wang, H. S. Li, R. L. Yin, J. Liu, Q. Dong, W. C. Wang, J. Yuan, B. S. Liu, X. H. Han, J. B. He and W. L. Bai, *Research in Veterinary Sci.,* 2016, 105, 65–73.
[13] Hau, K., T. H. Kwan and P. K. Li, *J. Am. Soc. Nephrol.,* 2009, 20, 245–250.
[14] Yin, R. H., X. Z. Wang, W. L. Bai, C. D. Wu, R. L. Yin and C. Li, *Res. Vet. Sci.,* 2013, 94, 618–627.

[15] Yoon, Y. S., D. H. Kim, S. K. Kim, S. B. Song, Y. Uh, D. Jin, X. F. Qi, Y. C. Teng and K. J. Lee, *Food Chem. Toxicol.*, 2011, 49, 1814–1819.

[16] Yin, R. H., X. Z. Wang, W. L. Bai, C. D. Wu, R. L. Yin and C. Li, *Res. Vet. Sci.*, 2013, 94, 618–627.

[17] Chang, L., R. She, I. Ma, H. You, F. Hu, T. Wang, X. Ding, Z. Guo and M. H. Soomro, *Reprod. Toxicol.*, 2014, 46, 1–11.

[18] Zhang, Q. X., G. Y. Yang, W. X. Li, B. Zhang and W. Zhu, *Regul. Toxicol. Pharmacol.*, 2011, 60, 144–150.

[19] Lin, M., L. He and J. Awika, *J. Food Sci.*, 2008, 73, T129–T134.

[20] Xia, X., S. Y. Ding, X. W. Li, X. Gong, S. X. Zhang, H. Y. Jiang, J. C. Li and J. Z. Shen, *Anal. Chim. Acta*, 2009, 651, 196–200.

[21] Xing, H. B., Y. G. Wu, S. S. Zhan and P. Zhou, *Food Anal. Method*, 2013, 6, 1–7.

[22] Filigenzi, M. S., B. Puschner, L. S. Aston and R. H. Poppenga, *J. Agric. Food Chem.*, 2008, 56, 7593–7599.

[23] Ehling, S., S. Tefera and I. P. Ho, *Food Addit. Contam.*, 2007, 24, 1319–1325.

[24] Xu, G., H. Zhang, M. Zhong, T. Zhang, X. Lu and X. Kan, *J. Electroanal. Chem.*, 2014, 713, 112–118.

[25] Xu, G., H. Zhang, M. Zhong, T. Zhang, X. Lu and X. Kan, *J. Electroanal. Chem.*, 2014, 713, 112–118.

[26] Li, N., T. Liu, S. G. Liu, S. M. Lin, Y. Z. Fan, H. Q. Luo and N. B. Li, *Sens. Actuators B,* 2017, 248, 597–604.

[27] Shen, J., Y. Yang, Y. Zhang, H. Yang, Z. Zhou and S. Yang, *Sens. Actuators B,* 2016, 226, 512–517.

[28] Rahman M. M. and J. Ahmed, *Biosens. Bioelectron,* 2018, 102, 631–636.

[29] Huy, B. T., Q. T. Pham, N. T. T. An, E. Conte and Y. I. Lee, *J. Lumin.*, 2017, 188, 436–440.

[30] Ezhilan, M., M. B. Gumpu, B. L. Ramachandra, N. Nesakumar, K. J. Babu, U. M. Krishnan and J. B. B. Rayappan, *Sens. Actuators B,* 2017, 238, 1283–1292.

[31] Rovina K. and S. Siddiquee, *Food Control,* 2016, 59, 801-808.

[32] Lee, K. S., C. W. Park and J. D. Kim, *Colloids and Surfaces A: Physicochem. Eng. Aspects*, 2017, 512, 87–92.

[33] Chakraborty, M., P. Mahapatra and R. Thangavel, *Thin Solid Films*, 2016, 612, 49–54.

[34] Zeng, H., Y. Cao, S. Xie, J. Yang, Z. Tang, X. Wang and L. Sun, *Nanoscale Res. Lett.*, 2013, 8, 133.

[35] Alam, M. M., A. M. Asiri, M. T. Uddin, M. A. Islam and M. M. Rahman, *ChemistrySelect*, 2018, 3, 11460–114.

[36] Rahman, M. M., M. M. Alam and A. M. Asiri, *Journal of Industrial and Engineering Chemistry*, 2018, 65, 300-308.

[37] Alam, M. M., A. M. Asiri, M. T. Uddin, M. A. Islam and M. M. Rahman, *RSC Adv.*, 2018, 8, 12562–12572.

[38] Rahman, M. M., M. M. Alam, A. M. Asiri and M. R. Awual, *New J. Chem.*, 2017, 41, 9159-9169.

[39] Rahman, M. M., M. M. Alam, A. M. Asiri and M. A. Islam, *RSC Adv.*, 2017, 7, 22627–22639.

[40] Rahman, M. M., M. M. Alam, A. M. Asiri and M. A. Islam, *Talanta*, 2017, 170, 215–223.

[41] Grigore, M. E., E. R. Biscu, A. M. Holban, M. C. Gestal and A. M. Grumezescu, *Pharmaceuticals*, 2016, 9, 75.

[42] Khan, P., A. Khan, M. M. Rahman, A. M. Asiri and M. Oves, *International Journal of Biological Macromolecules*, 2017, 98, 256–267.

[43] Rahman, M. M. *Sensors and Actuators B*, 2018, 264, 84–91.

[44] Rahman, M. M., M. M. Alam, M. M. Hussain, A. M. Asiri and M. E. M. Zayed, *Environmental Nanotechnology, Monitoring & Management, 2018*, 10, 1-9.

[45] Rahman, M. M., M. M. Alam and A. M. Asiri, *RSC Adv.*, 2018, 8, 960–970.

[46] Rahman, M. M., M. M. Alam, A. M. Asiri and M. A. Islam, *Talanta*, 2018, 176, 17–25.

[47] Speight, J. G. *McGraw-Hill*, 2005, 1, 1.331-1.342.

[48] Zhang, J., D. Gao, G. Yang, J. Zhang, Z. Shi, Z. Zhang, Z. Zhu and D. Xue, *Nanoscale Res. Letters*, 2011, 6, 587.

[49] Chiu F. C. and W. P. Chiang, *Mater.*, 2015, 8, 5795-5805.
[50] Bhowmick, T. K., A. K. Suresh, S. G. Kane, A. C. Joshi and J. R. Bellare, *J. Nanopart. Res.* 2009, 11, 655–664.
[51] Nakamura A. and J. Temmyo, *J. Appl. Phys.* 2011, 109, 093517.
[52] Dhara, S., K. Imakita, M. Mizuhata and M. Fujii, *Nanotechnology,* 2014, 25, 225202.
[53] Qamar, M. T., M. Aslam, Z. A. Rehan, M. T. Soomro, J. M. Basahi, I. M. I. Ismail, T. Almeelbi and A. Hameed, *Appl. Catal. B: Environ.*, 2017, 201, 105–118.
[54] Rahman, M. M., S. B. Khan, H. M. Marwani, A. M. Asiri and K. A. Alamry, *Chem. Central. J.* 2012, 6, 158.
[55] Peng, B., S. Zhang, S. Yang, H. Wang, H. Yu, S. Zhang and F. Peng, *Mater. Res. Bulletin,* 2014, 56, 19–24.
[56] Akhavan, O., R. Azimirad, S. Safad and E. Hasani, *J. Mater. Chem.,* 2011, 21, 9634–9640.
[57] Huang, M., Y. Zhang, F. Li, Z. Wang, Alamusi, N. Hu, Z. Wen and Q. Liu, *Scientific Report,* 2014, 4, 4518.
[58] Burriel, M., G. Garcia, J. Santiso, A. Abrutis, Z. Saltyte and A. Figueras, *Chem. Vap. Depos.,* 2005, 11, 106–111.
[59] Xia, H., D. Zhu, Z. Luo, Y. Yu, X. Shi, G. Yuan and J. Xie, *Scientific Report,* 2013, 3,) 2978.
[60] Sundar, L. S., G. O. Irurueta, E. V. Raman, M. K. Singh, A. C. M. Sousa, *Case Studies in Thermal Engineering,* 2016, 7, 66–77.
[61] Huang, Y., J. Chen, X. Zhang, Y. Zan, X. Wub, Z. He, H. Wang and Q. Li, *Chem. Engineer. J.,* 2016, 296 28–34.
[62] Chen, Z., S. Chen, Y. Li, X. Si, J. Huang, S. Massey and G. Chen, *Mater. Res. Bulletin,* 2014, 57, 170–176.
[63] Prabhu, Y. T., K. V. Rao, V. S. S. Kumar and B. S. Kumari, *Adv. Nanopart.,* 2013, 2, 45-50.
[64] Akhtar, M. J., M. Ahamed, S. Kumar, M. A. M. Khan, J. Ahmad and S. A. Alrokayan, *Inter. J. Nanomedicine,* 2012, 7, 845–857.
[65] Zak, K., R. Razali, W. H. A. Majid and M. Darroudi, *Inter. J. Nanomedicine,* 2011, 6, 1399–1403.

[66] Znaidi, L., T. Touam, D. Vrel, N. Souded, S. B. Yahia, O. Brinza, A. Fischer and A. Boudrioua, *Coatings,* 2013, 3, 126-139.
[67] Ashok, H., K. V. Rao and C. H. S. Chakra, *J. Atom. Molecul.,* 2014, 4, 803–806.
[68] Azam, A., A. S. Ahmed, M. Oves, M. S. Khan and A. Memic, *Inter. J. Nanomedicine,* 2012, 7, 3527–3535.
[69] Suleiman, M., M. Mousa and A. I. A. Hussein, *J. Mater. Environ. Sci.,* 2015, 6, 1924-1937.
[70] Kim K. S. and Y. J. Park, *Nanoscale Res. Letters,* 2012, 7, 47.
[71] Yoon T. H. and Y. J. Park, *Nanoscale Res. Letters,* 2012, 7, 28.
[72] Srivastava, R. *J. Sensor Technol.,* 2012, 2, 8-12.
[73] Mashford, B., J. Baldauf, T. L. Nguyen, A. M. Funston and P. Mulvaney, *J. Appl. Phys.* 2011, 109, 094305.
[74] Fooladsaz, K., M. Negahdary, G. Rahimi, A. H. Tamijani, S. Parsania, H. Akbari-dastjerdi, A. Sayad, A. Jamaleddini, F. Salahi and A. Asadi, *Int. J. Electrochem. Sci.,* 2012, 7, 9892 – 9908.
[75] Anzlovar, A., Z. C. Orel, K. Kogej and M. Zigon, *J. Nanomaterials,* 2012, 9, 760872.
[76] Arun, K. J., A. K. Batra, A. Krishna, K. Bhat, M. D. Aggarwal and P. J. J. Francis,. *Am. J. Mater. Sci.,* 2015, 5, 36-38.
[77] Alwan, R. M., Q. A. Kadhim, K. M. Sahan, R. A. Ali, R. J. Mahdi, N. A. Kassim and A. N. Jassim, *Nanosci. Nanotechnol.,* 2015, 5, 1-6.
[78] Cheng, J. P., X. Chen, R. Ma, F. Liu and X. B. Zhang, *Mater. Charact.,* 2011, 62, 775 – 780.
[79] Shah, H., E. Manikandan, M. B. Ahmed and V. Ganesan, *J. Nanomed. Nanotechol.,* 2013, 4, 3.
[80] Zhang D. and W. Zou, *Current Appl. Phys.,* 2013, 13, 1796-1800.
[81] Xu, H., Z. Hai, J. Diwu, Q. Zhang, L. Gao, D. Cui, J. Zang, J. Liu and C. Xue, *J. Nanomaterials,* 2015, 8, 845983.
[82] Ren, S., C. Li, X. Zhao, Z. Wu, S. Wang, G. Sun, Q. Xin and X. Yang, *J. Membrane Sci.,* 2005, 247, 59–63.
[83] Wang, Z., G. Liu, L. Zhang and H. Wang, *Ionics,* 2013, 19, 1687–1693.

[84] Rahman, M. M., M. M. Alam and A. M. Asiri, *New J. Chem.*, 2017, 41, 9938-9946.

[85] Rahman, M. M., M. M. Alam and A. M. Asiri, *RSC Adv.*, 2018, 8, 960–970

[86] Rahman, M. M., J. Ahmed and A. M. Asiri, *Sens. Actuators B,* 2017, 242, 167–175.

[87] Rahman, M. M., H. Balkhoyor and A. M. Asiri, *RSC Adv.*, 2016, 6, 29342-29352.

[88] Rahman, M. M., H. B. Balkhoyora and A. M. Asiri, *J. Environ. Management,* 2017, 188, 228-237.

[89] Rahman, M. M., H. B. Balkhoyor and A. M. Asiri, *J. Taiwan Institute of Chem. Engineers*, 2016, 66, 336-346.

[90] Balkhoyor, H. B., M. M. Rahman and A. M. Asiri, *RSC Adv.*, 2016, 6, 58236-58246.

[91] Maurino, V., M. Minella, F. Sordello and C. Minero, *Appl. Catalysis A: Gen.,* 2016, 521, 57–67.

[92] Piccinini, P., C. Minero, E. Pelizzetti and M. Vincenti, *J. Chem. Soc. Faraday Trans.,* 1997, 93, 1993–2000.

In: An Introduction to Melamine
Editor: Ashley Harris

ISBN: 978-1-53617-136-5
© 2020 Nova Science Publishers, Inc.

Chapter 5

FORMATION OF MELAMINE-DERIVED PARTICLES IN AQUEOUS AND BIOLOGICAL MATRICES

N. S. Chong[1,], PhD, D. Dutta[2], PhD and B. G. Ooi[1], PhD*

[1]Department of Chemistry, Middle Tennessee State University, Murfreesboro, Tennessee, US

[2]Department of Medicine, New York Medical College, Valhalla, New York, US

ABSTRACT

Melamine and cyanuric acid had been implicated in kidney-related diseases in infants and in the death of a large number of cats and dogs that ingested tainted food containing melamine. These incidents were caused by the willful adulteration of the raw ingredients with melamine in the dairy products and pet food, respectively, in order to boost the apparent protein content in the nutritional labels. Melamine and cyanuric acid can form extremely insoluble particles, which are composed of hydrogen-bonded melamine-cyanuric acid complex or melamine-cyanurate. Ingested

---

* Corresponding Author's Email: ngee.chong@mtsu.edu.

melamine and cyanuric acid are both absorbed in the gastrointestinal tract, distribute systemically, and precipitate as the melamine-cyanurate complex in the renal tubules, leading to progressive tubular blockage, degeneration, and acute renal failure. Melamine ingestion has also been implicated in neurological and reproductive toxicity. Due to the very active research and development of screening techniques aimed at reducing the incidence of melamine food contamination, the risk of renal failures due to melamine has been curtailed significantly. However, there are other routes of exposures to melamine and cyanuric acid that would still contribute to the adverse human health effects. The U. S. Food and Drug Administration (FDA) reported that melamine is incorporated into melamine-formaldehyde resins for making food packaging materials, plastic tableware, and the coating of cans in canned foods. Consequently, food and beverage products have been found to contain melamine at trace levels as a result of leaching from melamine-containing resins. Trace levels of cyanuric acid can be present in food and water from the use of dichloroisocyanurate in drinking water, swimming pools, and water used in food manufacturing. Cyanuric acid derivatives are also found in sanitizing solutions for food processing equipment and utensils.

This chapter will evaluate the formation of mealime-derived particles in aqueous and biological matrices using different analytical techniques for studying the bioaccumulation of melamine-cyanurate in tissues including kidney stones. The histomorphologic characteristics of the crystals formed at various concentrations and temperatures can be charatertized using scanning electron microscopy (SEM) to determine the crystallite morphology, size, and distribution. Our data indicated that the melamine-cyanurate crystals produced at 37°C were coarser and larger compared to those formed at 25°C at 100 ppm levels. Furthermore, the proportion of "spoke-like" crystals decreased along with the accompanying increase in the proportion of "needle-like" crystals at the higher temperature of 37°C. Both bovine serum albumin and polyvinylpyrrolidone, a synthetic macromolecule, have been found to alter the crystal morphology to a spherical form, which is typically observed for the particles in the kidney microtubules. Samples containing melamine-cyanurate formed in bovine blood plasma and in the kidney tissue of catfish that had been fed daily for 3 days with 200 milligram per day of melamine-cyanuric acid complex per kilogram of body weight were also analyzed by SEM and Raman microscopy. Other researchers have used X-ray diffraction (XRD), SEM with energy dispersive X-ray microanalysis, and Fourier transform IR spectroscopy to classify urinary stones. The results show that oxalates (43.3%) are the most common followed by phosphates (13.3%), urates (6.7%), and mixed stone (36.7%).

Keywords: melamin complex formation; melamine-related health effects; Melamine characterization methods; Sources of melamine contamination

INTRODUCTION

Melamine (2,4,6-triamino-1,3,5-triazine; $C_3H_6N_6$) consists of three reactive amine (-NH_2) groups attached to an aromatic s-triazine ring. It can be metabolized by microorganisms or bacteria to form analogues such as ammeline, ammelide, or cyanuric acid by replacing the amino groups with hydroxyl groups at each position as shown in Figure 1 (Wackett et al. 2002, 39-45). These analogues can also be produced by acid or base hydrolysis of melamine at high temperature. In fact, melamine and cyanuric acid can spontaneously aggregate in an aqueous environment to form stable, high molecular weight complexes, either through networks of intermolecular hydrogen bonds or π-π aromatic ring stacking (Seto and Whitesides 1993, 905-916). These complexes with extended lattice structure are shown in Figure 2. The stable melamine-cyanurate crystals are less soluble in water than either melamine or cyanuric acid.

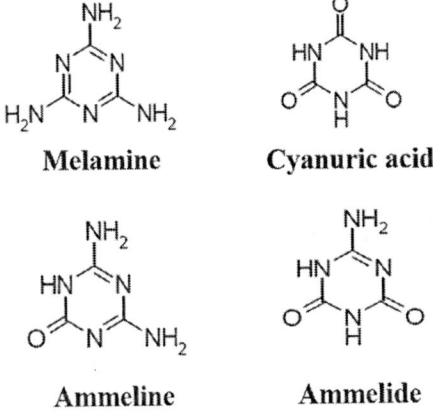

Figure 1. Structures of melamine, cyanuric acid, ammeline, and ammelide.

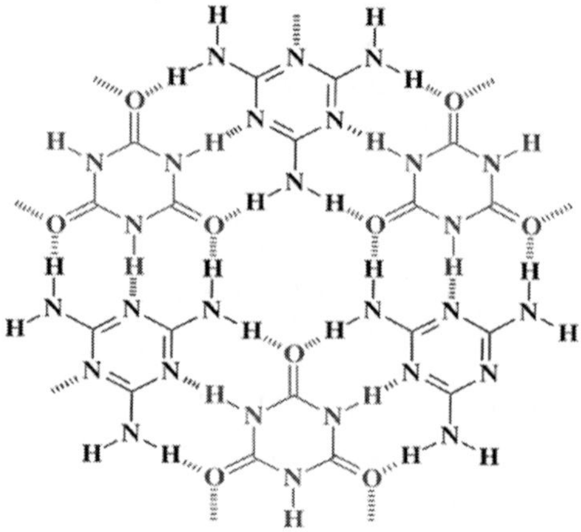

Figure 2. Melamine-cyanuric acid lattice structure. Melamine is shown in blue and cyanuric acid shown in red.

Table 1. Physical properties of melamine, cyanuric acid, ammeline, ammelide and melamine-cyanurate

	Melamine	Cyanuric acid	Ammeline	Ammelide	Melamine-cyanurate
Chemical formula	$C_3H_6N_6$	$C_3H_3N_3O_3$	$C_3H_5N_5O$	$C_3H_4N_4O_2$	$C_6H_9N_9O_3$
Molecular weight (g/mol)	126.12	129.07	127.10	128.09	255.19
Nitrogen % (w/w)	66.6	32.6	55.1	43.7	49.4
Melting point (°C)	345-347	360	Decomposes	Decomposes	350
Solubility in water (mg/L)	3240	2000	75	77	2.2
pKa	5.35	4.74	9.65	**	**

** = Not available in literature.

Melamine and its analogues are thermally stable and insoluble in many solvents. Physical properties of melamine and its analogues are shown in Table 1. Melamine is slightly soluble in water and alcohol, but it is more soluble at acidic pH (Bann and Miller 1958, 131-172; World Health Organization & Food and Agriculture Organization of the United Nations

2009). Cyanuric acid is less soluble than melamine in water, but is soluble in hot alcohols, concentrated H_2SO_4 and in alkaline pH. Ammelide has low solubility in water but is soluble in concentrated mineral acids and ammonia. Ammeline also has low solubility in water but is soluble in aqueous alkalis and mineral acids. Melamine-cyanurate has the lowest solubility in water (2.2 mg/L).

MELAMINE CONTAMINATION OF PET FOOD AND BABY FORMULA

In 2007 there was a large-scale pet food recalls in North America, Europe and South Africa because of crystal nephropathy and renal failure outbreak in cats and dogs (Puschner et al. 2007, 616-624). These pet foods had melamine contamination. The problem of melamine adulterated food became more pronounced in 2008 when over 294,000 children in mainland China and Hong Kong were hospitalized and treated for renal complications associated with ingestion of baby-food and formula tainted with melamine (Guan et al. 2009, 1067-1074; Ingelfinger 2008, 2745-2748). It was discovered that melamine, a nitrogen-rich organic compound, was deliberately added to pet and infant formula to inflate apparent protein content. According to the Institute of Nutrition and Food Safety, Chinese Center for Disease Control and Prevention, melamine levels in infant formula samples of Sanlu brand had a mean concentration of 1212 mg/kg (range <0.05 to 4700 mg/kg) (World Health Organization & Food and Agriculture Organization of the United Nations 2009). The contaminants identified in these adulterated foods were melamine (8.4%), cyanuric acid (5.3%), ammelide (2.3%), ammeline (1.7%), ureidomelamine and methylmelamine (both <1%) (Dobson et al. 2008, 251-262). While melamine and cyanuric acid alone at lower doses has low acute toxicity, excessive amount when co-ingested, results in melamine cyanurate crystal formation in kidneys (World Health Organization 2009). Infants and children are particularly vulnerable to these chemicals because melamine was shown to cross placenta and accumulate in the developing fetus in rats (Jingbin et al. 2010, 1791-1795).

Melamine is used in a large number of industrial and household products. It is reacted with formaldehyde to produce melamine resins for the manufacturing of glues, adhesives, laminates, plastics, coatings and flame retardants. Melamine is also present in some fertilizers and pesticides as the metabolite of cyromazine (Suchý et al. 2009, 55-59). Melamine is especially likely to assimilate into human tissues or organs through food. Cyanuric acid, a structural analogue and an impurity of melamine is present in biuret or carbamyl urea that is added to ruminant feed as a non-protein nitrogen source (Sprando et al. 2012, 4389-4397). Swimming pool water is also a good source of cyanuric acid exposure. Chlorinated isocyanurates such as dichloroisocyanurates used for cleansing, bleaching and disinfecting swimming pools hydrolyze in the presence of water to produce cyanurate and hypochlorous acid (Hammond et al. 1986, 287-292). Due to the rapid development and use of analytical techniques to screen food products for the presence of melamine and cyanuric acid, incidents of food contamination by melamine have not been reported in recent years. However, there are other routes of exposure to melamine and cyanuric acid that may still contribute to adverse human health effects. For instance, the U. S. Food and Drug Administration (FDA) reported that melamine is incorporated into melamine-formaldehyde resins for making food packaging materials, plastic tableware, and the coating of food tins but only residual amounts leach into food. The highest migration of melamine into 4% acetic acid, a food-simulating ingredient similar to vinegar used for cooking, was 42.9 ± 7.2 ppm when the migration test was repeated seven times at 95°C for 30 min. For the articles tested with distilled water, the melamine migration levels were found to have median concentrations of 22.2, 49.3, and 84.9 ng/ml at 25°C, 70°C, and 100°C, respectively (Ishiwata, Inoue, and Tanimura 1986, 63-69). Another study shows that in 3% acetic acid, melamine migration was at the levels of 31.5, 81.5, and 122.0 ng/ml at 25°C, 70°C, and 100°C, respectively (Chik et al. 2011, 967-973). The migration of melamine from plastic tableware into olive oil was not detectable and at least 20-fold lower compared to the aqueous food simulants. The migration levels of melamine into hot acidic beverages such as apple juice, tomato juice, and black coffee were rather similar to the acetic acid simulant (Bradley et al. 2010, 1755-1764).

Melamine and cyanuric acid have been traditionally tested in the laboratories with high sensitivity and accuracy using analytical chemistry techniques such as gas chromatography-mass spectrometry (GC-MS) and GC-MS-MS (Tzing and Ding 2010, 6267-6273; Zhao et al. 2010, 365-368), liquid chromatography with mass spectrometry *(LC-MS)* and LC-MS-MS (Tittlemier et al. 2009, 5340-5344; Yu et al. 2010, 48-58), capillary electrophoresis (Tsai et al. 2009, 8296-8303), and surface-enhanced Raman spectroscopy (SERS) (Jiang, Zhou, and Liang 2011, 3162-3164). Samples contaminated with melamine are often screened using immunoassays such as enzyme-linked immunosorbent assay (ELISA) (Zhou et al. 2012, 2681-2686). Another variation of immunoassay called lateral flow immunoassay (LFIA) has also been successfully used to screen for melamine during field applications (Le et al. 2013, 1610-1615). Currently, biosensors and chemosensors of various kinds such as electrochemical biosensors (Cao et al. 2009, 484-488), fluorescence chemosensors (Vasimalai and John 2013, 267-272), and electrochemi-luminescence chemosensors (Zhou et al. 2014, 57-63) have become popular for rapid melamine detection and screening because of its sensitivity, portability, and real-time monitoring advantages. Nanomaterial-based microfluidic devices have also developed for melamine detection (Zhai et al. 2010, 785-789).

TOXIC EFFECTS OF MELAMINE AND CYANURIC ACID

Toxicokinetics

Melamine is low in toxicity. When ingested, melamine will not be metabolized, but is rapidly absorbed and distributed in the body fluids. The half-life of melamine in plasma is approximately 3 hour and a vast majority of melamine is excreted through urine (World Health Organization 2009).

Due to the regular exposure to melamine containing compounds, the normal estimated oral uptake of melamine is approximately 0.007 mg melamine/kg/day (World Health Organization 2009). The US FDA has recommended 0.63 mg melamine/kg body weight/day as the tolerable daily intake (TDI). Melamine toxicity studies in rats suggested an oral median lethal dose (LD_{50}) of 1361 mg/kg body weight (World Health Organization 2009).

Unlike melamine, cyanuric acid has very few pharmacokinetic studies. It is not as widely used in consumer products and is more rapidly eliminated physiologically compared to melamine (Jacob et al. 2012, 317-324). In humans, within 24 hours approximately 98% of orally ingested cyanuric acid is excreted unmetabolized in urine (World Health Organization 2009). The TDI of 1.5 mg/kg body weight of cyanuric acid has been established by WHO (World Health Organization & Food and Agriculture Organization of the United Nations 2009). The oral LD_{50} in rats is 7700 mg/kg body weight (World Health Organization 2009). While cyanuric acid itself has low acute toxicity, sub-chronic exposure may cause renal damage and gastrointestinal track irritation (Hammond et al. 1986, 287-292).

In rats, the decomposition of melamine cyanurate into individual triazines is partial and slow as indicated by lower bioavailability for both melamine and cyanuric acid in serum (Jacob et al. 2012, 317-324). The elimination half-life of melamine cyanurate is also longer compared to melamine and cyanuric acid individually. This leads to the prolonged retention of melamine cyanurate in the body and triggers crystal formation in the kidneys. The LD_{50} of melamine cyanurate in rats is 4110 mg/kg body weight (Jacob et al. 2012, 317-324).

Renal Toxicity

In most animals, crystals of melamine cyanurate can readily form insoluble precipitates in the kidney by hydrogen bonding and pi-pi aromatic ring stacking (Zheng et al. 2013, 172ra22). It results in progressive nephron blockage and degeneration. Melamine-induced impaired kidney function was manifested by increased levels of blood urea nitrogen and creatinine in sheep where two third of them died after consuming 10 g melamine per day for 16-31 days (Hau, Kwan, and Li 2009, 245-250). Other forms of urinary pathology include crystalluria, proteinuria, microscopic hematuria and loss of specific gravity (Hau, Kwan, and Li 2009, 245-250). Melamine cyanurate crystals can form in both the proximal and distal tubules (Guan and Deng 2016, 613-617). Severe renal interstitial edema and hemorrhage at the corticomedullary junction in kidney cross-sections was also observed (Hau, Kwan, and Li 2009, 245-250). However, melamine-associated nephrolithiasis in infants was mostly composed of melamine and uric acid (Chang et al. 2012, 985-991). Imide groups in urea can interact with melamine at acidic pH to form melamine-urate crystals (Zheng et al. 2013, 172ra22). Thus, melamine ingestion can cause renal toxicity either due to melamine cyanurate, melamine-urate crystal or both. It has been reported that intestinal microbial composition plays a vital role in converting melamine into cyanuric acid and facilitate renal crystal formation (Zheng et al. 2013, 172ra22). *Klebsiella* sp. can deaminate melamine into cyanuric acid in the gastrointestinal tract. At normal urinary pH of about 6, melamine has a stronger affinity for hydrogen bonding with cyanuric acid whereas under acidic condition of pH 4, melamine exhibit a stronger affinity for uric acid instead. Hence, melamine cyanurate crystal formation takes precedence in the kidney under normal urinary pH condition. However, the progressive sedimentation of melamine cyanurate in renal tubules can promote further melamine-urate crystal formation that results in chronic renal toxicity (Zheng et al. 2013, 172ra22).

Developmental and Neurological Toxicity

Melamine exposure can lead to developmental and neurological defects. In rodents, melamine was shown to accumulate in the uterus (Sun et al. 2016a, 501-510). Since the size of melamine is smaller (126 g/mol) than placental vessels, it can translocate from placenta to the fetus (Chu et al. 2010, 398-402; Jingbin et al. 2010, 1791-1795). So, maternal exposure could result in transfer of melamine to amniotic fluid, fetal tissues and fetal circulation. During the critical periods of development, fetal exposure to melamine may have lasting impact on brain and neuronal tissues. The review by An and Sun (An and Sun 2017, 301-309) cited various *in vitro* studies on neuronal cells that indicated that melamine can trigger apoptosis after being exposed to 33 mg/mL melamine for 24 hours (Han et al. 2011, 65-71), inhibit proliferation of differentiated cells, disrupt metabolism (Wang et al. 2011, 571-576), and cause hyperpolarization and spontaneous neuron firing (Yang et al. 2011, 167-174). Melamine toxicity in developing brain can also occur through breastfeeding. Lactating female rats exposed to melamine showed increased accumulation of melamine in the milk that was transferred to the plasma of the suckling pups (Chu et al. 2010, 398-402). Melamine from plasma can cross the blood-brain barrier and enter into the brain (Wu et al. 2009, 7595-7601). Cognitive and behavioral studies have demonstrated that rats exposed to melamine during developmental periods could suffer from hippocampus and central nervous system (CNS) defects. These were manifested in the forms of poor cognitive flexibility, hemiparalysis and piloerection (An and Sun 2017, 301-309). Necroscopic evaluation revealed neuronal loss and necrotic neurons in the hippocampus of these rats.

Reproductive Toxicity

Accumulation of melamine in testes may induce male infertility. Mice exposed to melamine at the LD_{50} level had damaged testes and defective spermatogenesis (Huang et al. 2018, 345-352; Yin et al. 2013, 618-627). Lower serum testosterone and an imbalance of the hypothalamus-pituitary-testicular axis was reported after oral melamine ingestion in mice (Huang et al. 2018, 345-352; Sun et al. 2016b, 135-141). They also had decreased superoxide dismutase (SOD) activity and increase in free radicals in testes. In addition, the activity of succinate dehydrogenase (*SDH*) and *lactate dehydrogenase (LDH) in testes were markedly reduced resulting in energy insufficiency in spermatogenic cells (Huang et al. 2018, 345-352). These physiological deficits affected sperm production and maturation. Thus, testes of melamine exposed mice had significant increase in apoptotic cells, low sperm count, reduced sperm motility and deformed sperm tails.*

ANALYTICAL CHARACTERIZATION OF MELAMINE-DERIVED CRYSTALLITES

The investigation of the melamine contamination of pet food and baby formula incidents and the subsequent laboratory studies of calculi formed by melamine-cyanurate and melamine-urate are greatly aided by instrumental techniques based on LC-MS, X-ray diffraction, scanning electron microscopy with energy-dispersive X-ray analysis, as well as infrared and Raman microanalysis. The histomorphological characteristics of the melamine cyanurate crystals formed in aqueous solutions are examined using scanning electron microscopy (SEM). The morphology, size, and distribution of the crystals formed at the temperatures of 25°C and 37°C at the initial melamine and cyanuric acid concentrations of 100 ppm and 250 ppm are compared under the same microscopic magnification in Figure 3A-D. The melamine-cyanurate crystals produced at 37°C were coarser and larger compared to those formed at 25°C. Furthermore, the proportion of "spoke-like" crystals at 25°C shows a decrease along with the accompanying increase in the proportion of "needle-like" crystals at the higher temperature of 37°C. The larger crystal sizes at higher temperature may be due to the slower rate of formation and smaller number of initial nucleation sites. Similarly, lower concentration of 100 ppm complex gave rise to sparser distribution of crystal structures that are coarser and larger compared to crystals formed at the same temperature at higher initial concentration of 250 ppm for melamine and cyanuric acid.

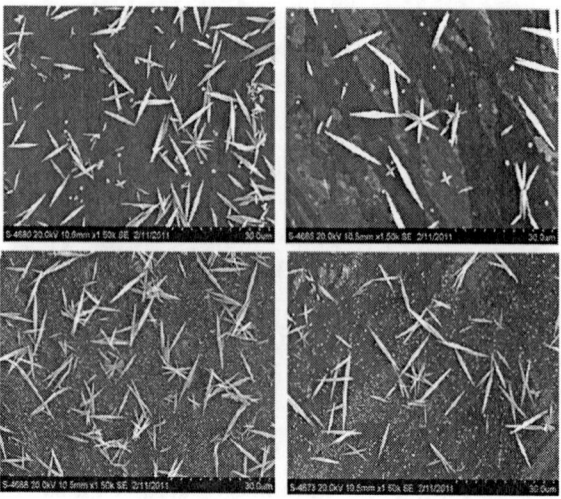

Figure 3. SEM of melamine-cyanuric acid complex formed from mixing 100 ppm and 250 ppm aqueous solutions of melamine and cyanuric acid separately. (A) Complex formed at 25°C at 100 ppmv; (B) Complex formed at 37°C at 100 ppmv; (C) Complex formed at 25°C at 250 ppmv; and (D) Complex formed at 37°C at 250 ppmv.

Morphology and Size of Melamine Complexes formed in Water and Blood Plasma

Cyanuric acid can exist in two tautomeric forms, lactam and lactim (EFSA Panel on Contaminants in the Food Chain (CONTAM) and EFSA Panel on Food Contact Materials, Enzymes, Flavourings and Processing Aids (CEF) 2010, 1573). The lactam form is stable as a solid, whereas the lactim form is stable in solution. When melamine interacts with cyanuric acid via hydrogen bonding to form melamine-cyanurate, the cyanuric acid is locked in the lactam tautomer and this makes melamine-cyanurate less soluble in water. This melamine-cyanurate form "needle-like" and "spoke-like" crystals in water. The rate of formation, amount, size, and crystal morphology are concentration and pH dependent. Higher concentrations increase the chances of the melamine and cyanuric interaction and therefore a large number of crystallites are formed quickly. Crystal growth is also accelerated at higher concentration so that most of the needle-like structures are complete, and several spoke-like structures are also observed. Both bovine serum albumin and polyvinylpyrrolidone, a synthetic macromolecule, have been found to alter the crystal morphology to a spherical form, which is typically observed for the particles in the kidney microtubules (Taksinoros and Murata 2013, 653-655).

Since crystal formation and growth are rapid at higher concentrations, the crystals have a fine "needle-like" morphology as opposed to coarser rod-like morphology of samples formed slowly. The average crystal length can range from around 36 μm at pH 8.8 to about 113 μm at pH 5.5. At pH 3.3 and 3.6, two crystal morphological forms were observed. One form had thick (17 μm width), smooth rod-like structure with blunt ends; whereas the other had much thinner (6.3 μm), furry stick-like structure (Akhter 2012).

Although melamine-cyanurate crystals were produced at pH conditions ranging from pH 3.3 to 8.8, only about 0.065 gram and 0.017 gram of crystals were formed at pH 3.3 and pH 8.8 respectively. There were significantly more crystals formed at pH 4.5, 5.5, and 6.5, with the highest amounts of 0.381 gram formed at pH 5.5 (Figure 4). This is because the rate of crystal formation reaches the maximum as the pH approaches 5.5, which is the optimal pH for the hydrogen bonding between melamine and cyanuric acid.

Melamine is a tribasic compound with the low pKa value of 5.1 and the high pKa value of 8.95 whereas cyanuric acid in solution is a triprotic compound with pKa values of 6.88, 11.4, and 13.5 (Tolleson et al. 2009). At pH below pH 4.5, melamine is converted from the unprotonated amine to the protonated ammonium form, which causes the destabilization of the hydrogen bonding with the lactam-form of cyanuric acid. At the same time, cyanuric acid at pH below 6.9 is protonated to the lactam-form favorable for hydrogen bonding with melamine, whereas above pH 6.9, cyanuric acid dissociates to the unprotonated charged form which does not favor hydrogen bonding with melamine. This explains why above pH 6.9 and below pH 4.5 melamine-cyanurate crystals formation via intermolecular hydrogen bonding was less likely to occur. At around pH 5.5, melamine exists predominantly in the unprotonated form and cyanuric acid exists in the lactam form. Since both of these forms are favorable for hydrogen bonding, the melamine-cyanurate crystals are formed more abundantly at around pH 5.5 compared to other pH conditions.

The pH of human blood ranges from pH 7.35 to 7.45, which is close to the bovine blood plasma that has a pH of 7.36. Formation of melamine-cyanurate complexes in bovine blood plasma had been studied at varying final reactant concentrations ranging from 0.1 mg/dL to 25 mg/dL (Akhter 2012). Below the concentrations of 0.7 mg/dL, no crystals were formed or detected using SEM. The crystals formed in the blood plasma were spherical and completely unlike the needle-like structures formed in water. Two sizes of crystals averaging about 1.8-4.1 µm and about 2.3-5.3 µm in diameter were observed at low concentrations of 0.7 mg/dL and 1 mg/dL respectively. At higher concentrations of 5 mg/dL, a denser formation and larger spherical crystals ranging from 3.4-7.3 µm in diameter, with the majority about 7.3 µm, were observed.

Figure 4. Comparison of melamine-cyanuate formed in water at different pH values. The reactant solutions were pH adjusted prior to mixing at 1:1 ratio and the final reactant concentration of the mixture were at 35 mg/dL. All aqueous mixtures were prepared at 37°C. Melamine-cyanurate purchased from Sigma-Aldrich and dissolved in water at 35 mg/dL was used as control. (A) pH 3.3; (B) pH 5.5; (C) pH 8.8; and (D) Control.

Raman Analysis of Melamine Complexes in Water and Blood Plasma

The SERS technique using the 25-fold pre-concentrated silver colloid prepared by the reduction of silver nitrate with hydroxylamine chloride (Leopold and Lendl 2003, 5723-5727) was found to be an effective method for distinguishing between melamine, cyanuric acid, uric acid, melamine-cyanurate and melamine-urate. The SERS method could detect melamine-cyanurate dissolved in blood plasma at the melamine and cyanuric acid concentrations as low as 2.5 mg/L, and melamine-cyanurate in water at the detection limit of 1 mg/L. The SERS spectra for melamine-cyanurate formed in water shows a prominent peak at 711.3 cm^{-1} along with other characteristic SERS peaks at 1077.4 cm^{-1} and 1245.8 cm^{-1} as shown in Figure 5 (Akhter 2012). The spectrum of melamine-cyanurate formed in blood plasma has a lower signal-to-noise ratio than that formed in aqueous solutions but still included the merged peaks at 711.3 cm^{-1} as well as the other characteristic peaks (Figure 5). The SERS technique offers the advantage of being able to analyze the aqueous samples and blood plasma with minimal sample preparation. However, the SERS detection limits are generally lower than those based on mass spectrometry.

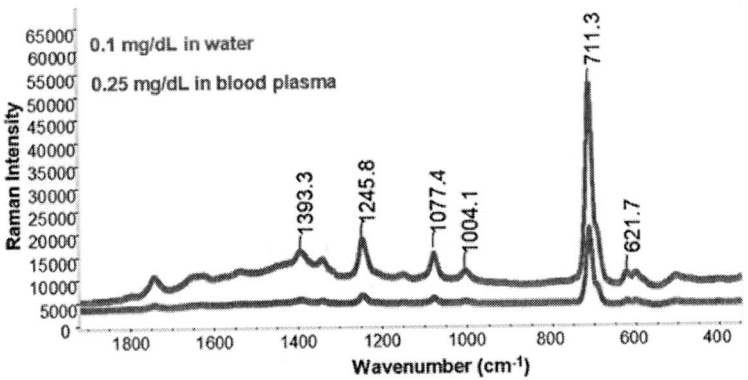

Figure 5. SERS spectra of melamine-cyanurate formed in water and blood plasma. The melamine-cyanurate in water and in blood plasma were produced at final reactant concentrations of 0.1 mg/dL and 0.25 mg/dL respectively.

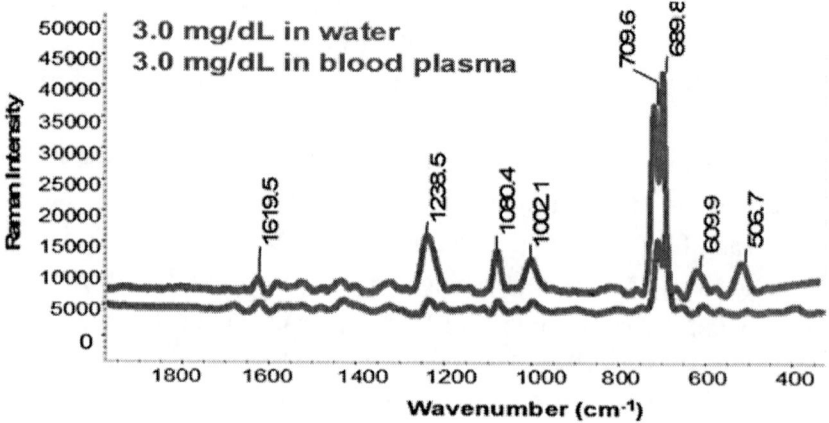

Figure 6. SERS spectra of melamine-urate formed in water and in blood plasma. The melamine-urate in water pH 5.5 and in blood plasma pH 7.36 were formed at final urate concentration of 3.0 mg/dL

Infants affected in the 2008 milk adulteration incident appeared to have developed calculi primarily in the urinary tract, which led to obstructive renal failure (Guan et al. 2009, 1067-1074). The kidney stones or calculi are commonly composed of melamine and uric acid instead of melamine and cyanuric acid. However, the histological examination of the kidney section could not detect any melamine-urate, probably because routine processing of the kidney section may have dissolved any melamine-urate crystals in the renal tubules (Vernon 2006, 17-19).

Raman spectroscopy provides a simple and rapid method for obtaining fingerprint-like spectra specific for the compound. Both the spectra of melamine-urate formed in water and in blood plasma showed the distinguishable double peaks at 709.6 and 689.8 cm^{-1}, as well as other characteristic peaks at 506.7, 1003.2, 1080.7, and 1238.5 cm^{-1}. These spectra were obtained for complex formed at final urate concentration as low as 3.0 mg/dL in water at pH4.5 and pH5.5 as well as in blood plasma at pH 7.36 (Figure 6).

CONCLUSION

Due to the unfortunate incidents of pet food and baby formula contamination with melamine, many scientific studies have been conducted that improve our understanding of the pathology of renal failures due to the formation of melamine-based kidney stones and further enhance the ability to screen for the presence of melamine in food and beverage products to avoid human exposure. Although the probability of intentional adulteration of food products with melamine is low now-a-days as a result of the awareness of consumers and the scrutiny by governmental agencies such as FDA, there is a distinct possibility that melamine exposure may still occur. Migration of melamine into food and drinks under heated and acidic conditions has been reported to occur from melamine-based dinnerware items. It is also important to point out that even though this chapter emphasizes the adverse effects of melamine-related kidney stones, other effects including reproductive, neurological, and bladder cancer are also important and need to be further investigated in the future.

REFERENCES

Akhter, Farhana. 2012. "*Toxicological Study of the Formation of Melamine-Cyanurate and Melamine-Urate Complexes in Aqueous and Physiological Matrices.*" MS Thesis, Middle Tennessee State University.

An, Lei and Wei Sun. 2017. "A Brief Review of Neurotoxicity Induced by Melamine." *Neurotoxicity Research* 32 (2): 301-309.

Bann, Bernard and Samuel A. Miller. 1958. "Melamine and Derivatives of Melamine." *Chemical Reviews* 58 (1): 131-172.

Bradley, EL, Laurence Castle, JS Day, Ingo Ebner, Karl Ehlert, Ruediger Helling, Sander Koster, Jennifer Leak, and Karla Pfaff. 2010. "Comparison of the Migration of Melamine from Melamine–formaldehyde Plastics ('melaware') into various Food Simulants and

Foods Themselves." *Food Additives & Contaminants: Part A* 27 (12): 1755-1764.

Cao, Qian, Hong Zhao, Lixi Zeng, Jian Wang, Rui Wang, Xiaohui Qiu, and Yujian He. 2009. "Electrochemical Determination of Melamine using Oligonucleotides Modified Gold Electrodes." *Talanta* 80 (2): 484-488.

Chang, Hong, Xiaofeng Shi, Wei Shen, Wei Wang, and Zhongjin Yue. 2012. "Characterization of Melamine-Associated Urinary Stones in Children with Consumption of Melamine-Contaminated Infant Formula." *Clinica Chimica Acta* 413 (11-12): 985-991.

Chik, Z., DE Mohamad Haron, ED Ahmad, H. Taha, and AM Mustafa. 2011. "Analysis of Melamine Migration from Melamine Food Contact Articles." *Food Additives & Contaminants: Part A* 28 (7): 967-973.

Chu, CY, KO Chu, Judy YW Chan, XZ Liu, CS Ho, CK Wong, CM Lau, TL Ting, TF Fok, and KP Fung. 2010. "Distribution of Melamine in Rat Foetuses and Neonates." *Toxicology Letters* 199 (3): 398-402.

Dobson, Roy LM, Safa Motlagh, Mike Quijano, R. Thomas Cambron, Timothy R. Baker, Aletha M. Pullen, Brian T. Regg, Adrienne S. Bigalow-Kern, Thomas Vennard, and Andrew Fix. 2008. "Identification and Characterization of Toxicity of Contaminants in Pet Food Leading to an Outbreak of Renal Toxicity in Cats and Dogs." *Toxicological Sciences* 106 (1): 251-262.

EFSA Panel on Contaminants in the Food Chain (CONTAM) and EFSA Panel on Food Contact Materials, Enzymes, Flavourings and Processing Aids (CEF). 2010. "Scientific Opinion on Melamine in Food and Feed." *EFSA Journal* 8 (4): 1573.

Guan, Na, Qingfeng Fan, Jie Ding, Yiming Zhao, Jingqiao Lu, Yi Ai, Guobin Xu, Sainan Zhu, Chen Yao, and Lina Jiang. 2009. "Melamine-Contaminated Powdered Formula and Urolithiasis in Young Children." *New England Journal of Medicine* 360 (11): 1067-1074.

Guan, Xiaofeng and Yaoliang Deng. 2016. "Melamine-Associated Urinary Stone." *International Journal of Surgery* 36: 613-617.

Hammond, B. G., S. J. Barbee, T. Inoue, N. Ishida, G. J. Levinskas, M. W. Stevens, A. G. Wheeler, and T. Cascieri. 1986. "A Review of

Toxicology Studies on Cyanurate and its Chlorinated Derivatives." *Environmental Health Perspectives* 69: 287-292.

Han, Yang-guang, Shi-chang Liu, Tao Zhang, and Zhuo Yang. 2011. "Induction of Apoptosis by Melamine in Differentiated PC12 Cells." *Cellular and Molecular Neurobiology* 31 (1): 65-71.

Hau, A. K., T. H. Kwan, and P. K. Li. 2009. "Melamine Toxicity and the Kidney." *Journal of the American Society of Nephrology* 20 (2): 245-250.

Huang, Jian, Guozhen Yang, Fengqiong Xia, and Shu Zhang. 2018. "Reproductive Toxicity of Melamine Against Male Mice and the Related Mechanism." *Toxicology Mechanisms and Methods* 28 (5): 345-352.

Ingelfinger, Julie R. 2008. "Melamine and the Global Implications of Food Contamination." *New England Journal of Medicine* 359 (26): 2745-2748.

Ishiwata, Hajimu, Takiko Inoue, and Akio Tanimura. 1986. "Migration of Melamine and Formaldehyde from Tableware made of Melamine Resin." *Food Additives & Contaminants* 3 (1): 63-69.

Jacob, Cristina C., Linda S. von Tungeln, Michelle Vanlandingham, Frederick A. Beland, and Gonçalo Gamboa da Costa. 2012. "Pharmacokinetics of Melamine and Cyanuric Acid and their Combinations in F344 Rats." *Toxicological Sciences* 126 (2): 317-324.

Jiang, Zhiliang, Lianping Zhou, and Aihui Liang. 2011. "Resonance Scattering Detection of Trace Melamine using Aptamer-Modified Nanosilver Probe as Catalyst without Separation of its Aggregations." *Chemical Communications* 47 (11): 3162-3164.

Jingbin, Wang, Moussa Ndong, Hisahiro Kai, Koji Matsuno, and Fujio Kayama. 2010. "Placental Transfer of Melamine and its Effects on Rat Dams and Fetuses." *Food and Chemical Toxicology* 48 (7): 1791-1795.

Le, Tao, Peifeng Yan, Jian Xu, and Youjing Hao. 2013. "A Novel Colloidal Gold-Based Lateral Flow Immunoassay for Rapid Simultaneous Detection of Cyromazine and Melamine in Foods of Animal Origin." *Food Chemistry* 138 (2-3): 1610-1615.

Leopold, Nicolae and Bernhard Lendl. 2003. "A New Method for Fast Preparation of Highly Surface-Enhanced Raman Scattering (SERS) Active Silver Colloids at Room Temperature by Reduction of Silver Nitrate with Hydroxylamine Hydrochloride." *The Journal of Physical Chemistry B* 107 (24): 5723-5727.

Puschner, Birgit, Robert H. Poppenga, Linda J. Lowenstine, Michael S. Filigenzi, and Patricia A. Pesavento. 2007. "Assessment of Melamine and Cyanuric Acid Toxicity in Cats." *Journal of Veterinary Diagnostic Investigation* 19 (6): 616-624.

Seto, Christopher T. and George M. Whitesides. 1993. "Molecular Self-Assembly through Hydrogen Bonding: Supramolecular Aggregates Based on the Cyanuric Acid-Melamine Lattice." *Journal of the American Chemical Society* 115 (3): 905-916.

Sprando, Robert L., Renate Reimschuessel, Cynthia B. Stine, Thomas Black, Nicholas Olejnik, Michael Scott, Zachary Keltner, Omari Bandele, Martine Ferguson, and Sarah M. Nemser. 2012. "Timing and Route of Exposure Affects Crystal Formation in Melamine and Cyanuric Exposed Male and Female Rats: Gavage Vs. Feeding." *Food and Chemical Toxicology* 50 (12): 4389-4397.

Suchý, Pavel, Eva Straková, Ivan Herzig, Jaroslav Staňa, Renata Kalusová, and Markéta Pospíchalová. 2009. "Toxicological Risk of Melamine and Cyanuric Acid in Food and Feed." *Interdisciplinary Toxicology* 2 (2): 55-59.

Sun, Huiying, Kaizhong Wang, Haiyan Wei, Zhe Li, and Hui Zhao. 2016a. "Cytotoxicity, Organ Distribution and Morphological Effects of Melamine and Cyanuric Acid in Rats." *Toxicology Mechanisms and Methods* 26 (7): 501-510.

Sun, Jiarui, Yinan Cao, Xinchen Zhang, Qiling Zhao, Endong Bao, and Yingjun Lv. 2016b. "Melamine Negatively Affects Testosterone Synthesis in Mice." *Research in Veterinary Science* 109: 135-141.

Taksinoros, Sarawut and Hideo Murata. 2013. "Effects of Polyvinylpyrrolidone on in Vitro Melamine-Cyanurate Crystal Formation: An Electron Microscopy Study." *Journal of Veterinary Medical Science* 75 (5): 653-655.

Tittlemier, Sheryl A., Benjamin P-Y Lau, Cathie Menard, Catherine Corrigan, Melissa Sparling, Dean Gaertner, Karen Pepper, and Mark Feeley. 2009. "Melamine in Infant Formula Sold in Canada: Occurrence and Risk Assessment." *Journal of Agricultural and Food Chemistry* 57 (12): 5340-5344.

Tolleson, W. H., G. W. Diachenko, D. Folmer, D. Doell, and D. Heller. 2009. "Background Paper on the Chemistry of Melamine Alone and in Combination with Related Compounds." *World Health Organization*, Geneva.

Tsai, I-Lin, Shao-Wen Sun, Hsiao-Wei Liao, Shu-Chiao Lin, and Ching-Hua Kuo. 2009. "Rapid Analysis of Melamine in Infant Formula by Sweeping-Micellar Electrokinetic Chromatography." *Journal of Chromatography A* 1216 (47): 8296-8303.

Tzing, Shin-Hwa and Wang-Hsien Ding. 2010. "Determination of Melamine and Cyanuric Acid in Powdered Milk using Injection-Port Derivatization and Gas Chromatography–tandem Mass Spectrometry with Furan Chemical Ionization." *Journal of Chromatography A* 1217 (40): 6267-6273.

Vasimalai, N. and S. Abraham John. 2013. "Picomolar Melamine Enhanced the Fluorescence of Gold Nanoparticles: Spectrofluorimetric Determination of Melamine in Milk and Infant Formulas using Functionalized Triazole Capped Goldnanoparticles." *Biosensors and Bioelectronics* 42: 267-272.

Vernon, Stephen E. 2006. "Preservation of Tissue Urate Crystals with the use of a Rapid Tissue-Processing System." *Journal of Histotechnology* 29 (1): 17-19.

Wackett, L., M. Sadowsky, Betsy Martinez, and Nir Shapir. 2002. "Biodegradation of Atrazine and Related s-Triazine Compounds: From Enzymes to Field Studies." *Applied Microbiology and Biotechnology* 58 (1): 39-45.

Wang, Yan, Fei Liu, Yuejiao Wei, and Daicheng Liu. 2011. "The Effect of Exogenous Melamine on Rat Hippocampal Neurons." *Toxicology and Industrial Health* 27 (6): 571-576.

World Health Organization. "*Toxicological and Health Aspects of Melamine and Cyanuric Acid.*" Accessed August, 2019, https://www.who.int/foodsafety/publications/melamine-cyanuric-acid/en/.

World Health Organization & Food and Agriculture Organization of the United Nations. "*Toxicological and Health Aspects of Melamine and Cyanuric Acid: Report of a WHO Expert Meeting in Collaboration with FAO, Supported by Health Canada, Ottawa, Canada, 1-4 December 2008. World Health Organization.*". Accessed August, 2019, https://apps.who.int/iris/handle/10665/44106.

Wu, Yu-Tse, Chih-Min Huang, Chia-Chun Lin, Wei-An Ho, Lie-Chwen Lin, Ting-Fang Chiu, Der-Cherng Tarng, Chi-Hung Lin, and Tung-Hu Tsai. 2009. "Determination of Melamine in Rat Plasma, Liver, Kidney, Spleen, Bladder and Brain by Liquid Chromatography–tandem Mass Spectrometry." *Journal of Chromatography A* 1216 (44): 7595-7601.

Yang, Jiajia, Lei An, Yang Yao, Zhuo Yang, and Tao Zhang. 2011. "Melamine Impairs Spatial Cognition and Hippocampal Synaptic Plasticity by Presynaptic Inhibition of Glutamatergic Transmission in Infant Rats." *Toxicology* 289 (2-3): 167-174.

Yin, Rong H., Xin Z. Wang, Wen L. Bai, Chang D. Wu, Rong L. Yin, Chang Li, Jiao Liu, Bao S. Liu, and Jian B. He. 2013. "The Reproductive Toxicity of Melamine in the Absence and Presence of Cyanuric Acid in Male Mice." *Research in Veterinary Science* 94 (3): 618-627.

Yu, Huan, Yanfei Tao, Dongmei Chen, Yulian Wang, Zhaoying Liu, Yuanhu Pan, Lingli Huang, Dapeng Peng, Menghong Dai, and Zhenli Liu. 2010. "Development of a High Performance Liquid Chromatography Method and a Liquid Chromatography–tandem Mass Spectrometry Method with Pressurized Liquid Extraction for Simultaneous Quantification and Confirmation of Cyromazine, Melamine and its Metabolites in Foods of Animal Origin." *Analytica Chimica Acta* 682 (1-2): 48-58.

Zhai, Chun, Wei Qiang, Jin Sheng, Jianping Lei, and Huangxian Ju. 2010. "Pretreatment-Free Fast Ultraviolet Detection of Melamine in Milk Products with a Disposable Microfluidic Device." *Journal of Chromatography A* 1217 (5): 785-789.

Zhao, Li, Jian Zhao, Wei-Guo Huangfu, and Yin-Liang Wu. 2010. "Simultaneous Determination of Melamine and Clenbuterol in Animal Feeds by GC–MS." *Chromatographia* 72 (3-4): 365-368.

Zheng, X., A. Zhao, G. Xie, Y. Chi, L. Zhao, H. Li, C. Wang, et al. 2013. "Melamine-Induced Renal Toxicity is Mediated by the Gut Microbiota." *Science Translational Medicine* 5 (172): 172ra22.

Zhou, Limin, Jianshe Huang, Lu Yang, Libo Li, and Tianyan You. 2014. "Enhanced Electrochemiluminescence Based on Ru (Bpy) 32 -Doped Silica Nanoparticles and Graphene Composite for Analysis of Melamine in Milk." *Analytica Chimica Acta* 824: 57-63.

Zhou, Yu, Chun-Yuan Li, Yan-Song Li, Hong-Lin Ren, Shi-Ying Lu, Xiang-Li Tian, Ya-Ming Hao, Yuan-Yuan Zhang, Qing-Feng Shen, and Zeng-Shan Liu. 2012. "Monoclonal Antibody Based Inhibition ELISA as a New Tool for the Analysis of Melamine in Milk and Pet Food Samples." *Food Chemistry* 135 (4): 2681-2686.

BIOGRAPHICAL SKETCHES

Beng Guat Ooi

Affiliation: Department of Chemistry, Middle Tennessee State University

Education:

1983 – Monash University, Australia B.S. in Biochemistry & Microbiology
1987– Monash University, Australia, Ph.D. in Biochemistry (Yeast mtDNA)
1987-1989 – University of Georgia, Research Associate in virology
1989-1990 – Baylor College of Medicine, Research Associate in human genetics

Business Address: 1301 East Main Street, Murfreesboro, TN 37132, USA

Research and Professional Experience:

Professor of Biochemistry specializing in the study of how yeasts and fungi metabolize chemicals such as sugars, hemicellulose, lignocellulose, and polycyclic aromatic hydrocarbons. Application of biological and spectroscopic techniques such as PCR, DNA sequencing, GC-MS, LC-MS, SERS, and FTIR methods for probing the interaction of microbes with chemical systems.

Professional Appointments:

Present – Full Professor, Middle Tennessee State University, Dept. of Chemistry
2005-2012 – Associate Professor, Middle Tennessee State University, Dept. of Chemistry
1999-2005 – Assistant Professor, Middle Tennessee State University, Dept. of Chemistry

Honors: Excellence In Publications - MTSU College of Basic and Applied Sciences 2008-2009.

Publications from the Last 3 Years:

1. B.G. Ooi, D. Dutta, K. Kazipeta, N.S. Chong. Influence of E-cigarette emission profile by the ratio of glycerol to propylene glycol in e-liquid composition. *ACS Omega* 2019, *4*, 13338-13348.
2. Y. E. Ejorh, W. H. Ilsley, B. G. Ooi. Elucidating the chemisorption phenomena in SERS studies via computational modeling. *Optics and Photonics Journal* 2018, *8*, 212-234.
3. B. G. Ooi and S. A. Branning. Correlation of conformational changes and protein degradation with loss of lysozyme activity due to chlorine dioxide treatment. *Applied Biochemistry and Biotechnology* 2017, 182(2), 782-791.

Ngee Sing Chong

Affiliation: Department of Chemistry, Middle Tennessee State University (MTSU)

Education:

1981 – Hanover College   B.A. in Chemistry
1985 – Iowa State University   M.S. in Analytical Chemistry
1991 – University of Georgia   Ph.D. Analytical/Solid State Chemistry
1989-1990 – Rice University   Welch Postdoctoral Research Fellow

Business Address: 1301 East Main Street, Murfreesboro, TN 37132, USA

Research and Professional Experience:

Professor of Chemistry and Director of MTSU Interdisciplinary Microanalysis & Imaging Center. His research group investigates analytical methods development for environmental, bioanalytical, forensic, and materials applications with research interests that range from the analysis of pollutants to the characterization of nanomaterials. The methods used in his research include chromatography, mass spectrometry, optical spectroscopy, and electron microscopy.

Professional Appointments:

2009-present – Full Professor, Middle Tennessee State University
2007-present – Director, MTSU Interdisciplinary Microanalysis & Imaging Center
2004-2009 – Associate Professor, Middle Tennessee State University
2004 & 2006 – Faculty Research Fellow, Oak Ridge National Laboratory
2004-2008 – Board Member, Tennessee Air Pollution Control Board
1998-2004 – Assistant Professor, Middle Tennessee State University
Summer 2002 – Summer Faculty Fellow, Air Force Research Laboratory

1994-1998 – Senior Chemist, Texas Commission on Environmental Quality
1991-1994 – Director of Research, Cantrell Research Incorporated

Honors: MTSU Distinguished Research Awards in 2000-2001 and 2013-2014; MTSU College of Basic and Applied Sciences Overall Excellence Award in 2007-2008.

Publications from the Last 3 Years:

1. B.G. Ooi, D. Dutta, K. Kazipeta, N.S. Chong. Influence of E-cigarette emission profile by the ratio of glycerol to propylene glycol in e-liquid composition. *ACS Omega* 2019, 4, 13338-13348.
2. O.A. Oladipupo, D. Dutta, N.S. Chong Analysis of chemical constituents in mainstream bidi smoke; *BMC Chemistry*, 2019, 13:93.
3. D. Dutta, N. S. Chong, S. H. Lim "Endogenous volatile organic compounds in acute myeloid leukemia: Origins and clinical applications" *Journal of Breath Research*, 2018, 12, 034002, https://doi.org/10.1088/1752-7163/aab108.
4. P. Villa, L. Vera, S. Wylie, S. Wilson, A. Septoff, C. Jia, N. S. Chong, C. Luong "Hazards in the Air: Relating reported illnesses to air pollutants detected near oil and gas operations in and around Karnes County, Texas" *Earthworks Report*, 2017, 96 pages.

Dibyendu Dutta

Affiliation: Department of Medicine, New York Medical College (NYMC)

Education:

1999 – University of Calcutta, India   B.Sc. in Botany (Hons.)
2001 – University of North Bengal, India   M.Sc. in Botany (Cytogenetics)
2011 – Texas Woman's University   Ph.D. Molecular Biology
2018 – Middle Tennessee State University   MSPS Health Care Informatics

Business Address: 15 Dana Road, Valhalla, NY 10595, USA

Research and Professional Experience:

Scientist at New York Medical College. His research is to investigate clinical significance and mechanism of how intestinal microbiota contributes to vaso-occlusive crisis in sickle cell disease.

Professional Appointments:

2018-present – Scientist, New York Medical College, Valhalla, New York
2013-2014 – Postdoctoral Fellow, Institute for Stem Cell Science and Regenerative Medicine, India
2012-2013 – Postdoctoral Researcher II, University of Texas Southwestern Medical Center, Dallas, Texas

Honors: Advancing Science Conference Grant, NOBCChE (2017 *and 2018);* International Merit Scholarship, Middle Tennessee State University (2016); Scholarship for Frontiers in Stem Cells and Regeneration, Marine Biological Laboratory (2011); Chancellor's Research Scholar Award, Texas Woman's University (2009 and 2011).

Publications from the Last 3 Years (partial list):

1. Dutta, Dibyendu, et al. "Intestinal injury and gut permeability in sickle cell disease." *Journal of Translational Medicine* 17.1 (2019): 183.
2. Ooi, Beng G., et al. "Influence of the E-cigarette emission profile by the ratio of glycerol to propylene glycol in e-liquid composition." *ACS Omega* 4.8 (2019): 13338-13348.
3. Dutta, Dibyendu, et al. "Ethylene dimethane sulfonate (EDS) ablation of Leydig cells in adult rat depletes testosterone resulting in epididymal sperm granuloma: Testosterone replacement prevents granuloma formation." *Reproductive Biology* 19.1 (2019): 89-99.

4. Oladipupo, Omobola Ajoke, Dibyendu Dutta, and Ngee Sing Chong. "Analysis of chemical constituents in mainstream bidi smoke." *BMC Chemistry* 13.1 (2019): 93.
5. Dutta, Dibyendu, et al. "Effects of rifaximin on circulating aged neutrophils in sickle cell disease." *American Journal of Hematology* 94.6 (2019): E175-E176.
6. Dutta, Dibyendu, Ngee S. Chong, and Seah H. Lim. "Endogenous volatile organic compounds in acute myeloid leukemia: origins and potential clinical applications." *Journal of Breath Research* 12.3 (2018): 034002.

In: An Introduction to Melamine
Editor: Ashley Harris

ISBN: 978-1-53617-136-5
© 2020 Nova Science Publishers, Inc.

Chapter 6

MELAMINE DETECTION WITH NANOSTRUCTURE MATERIALS

Mohammed Muzibur Rahman[*], PhD
Department of Chemistry, Faculty of Science,
King Abdulaziz University, Jeddah 21589, Saudi Arabia

ABSTRACT

In this approach, nanostructure materials (Cadmium doped antimony oxide; CAO NSs) were synthesized by a facile wet-chemical method at a low temperature to detect melamine from aqueous solution. The calcined CAO-NSs were characterized systematically by FE-SEM, EDS, UV/Vis., FTIR spectroscopy, powder XRD, and XPS techniques. The glassy carbon electrode (GCE) was modified with the CAO-NSs and sensing performance towards the selective melamine was explored by the electrochemical approach in phosphate buffer solution. The melamine undergoes a reduction reaction in the presence of CAO-NSs/GCE in PBS. The CAO-NSs/GC electrode attained higher sensitivity for a wide range of concentration, ultra-low limit of detection, long-term stability, excellent

---

[*] Corresponding Author's Email: mmrahman@kau.edu.sa, mmrahmanh@gmail.com, Fax: +966-12-695-2292.

repeatability, and reproducibility. This method might represent an efficient way of sensitive sensor development for the determination of toxic melamine and their derivatives in broad scales.

INTRODUCTION

Here, generally melamine (1,3,5-triazine-2,4,6-triamine) is a nitrogen rich (66% by mass) heterocyclic compound. It is commonly used to produce plastics, adhesive, coating-agents, dish-ware, foams, pigments, glues and flame-retardants etc. (Rovina et al. 2016; De-Araujo et al. 2014; Shen et al. 2017; Liu et al. 2015; Chen et al. 2015). Because of the high nitrogen-content, some fraudulent companies' used to use melamine in dairy products to show the high protein-contents in their products. It also increases the nitrogen concentration in milk producing a false appearance of a higher level- protein by the Kjeldahl method (Yang et al. 2009). The US Food and Drug Administration (FDA) and the European Union (EU) have set up maximum limits of melamine residue that can be present in different food items. Maximum limits of melamine content for infant-formula is 1.0 ppm and while 2.5 ppm for the rest of the dairy products have been enacted in many countries (Liu et al. 2015; Chen et al. 2009). Consumption of melamine beyond the authorized safety limit can cause cancer, kidney-damage, and malfunctions in the reproductive system especially in babies and children (Guan et al. 2009). For instance, the discovery of acute kidney problems to children and pets instigated extensive dairy and pet-food recall in USA and China in 2007-08 (Choi et al. 2016). In spite of strict rules, some food products still contain melamine higher than permissible levels (Li et al. 2015). Hence, a simple, sensitive, and robust method for detecting the melamine-level in dairy products is necessary to control this situation, which has become a global health concern (Sun et al. 2010; Chan et al. 2008).

So far, existing methods for melamine determination includes: spectrophotometry (Hirt et al. 1954), chemiluminescence (Sun et al. 2010; Zhang et al. 2011), electrochemistry (Wu et al. 2012; Zhu et al. 2010; Pietrzyk et al. 2009), chromatography combine with UV or mass detector

(Pan et al. 2013; Yu et al. 2010). However, reported compounds readily interfere in spectrophotometric determination. Meanwhile, chromatography requires an excessive volume of extra-pure organic solvent, sophisticated instrument, and expert hand. Furthermore, it is slow, problematic, and costly, and hence inappropriate for routine field analysis. Even though some of the existing methods show excellent performance in terms of sensitivity and limits of detection (LOD), virtually all require pre-treatment of analytes and multi-step reactions that restricts analytical application. However, some of the existing methods have been extremely effective in decreasing the assay-time, but are slightly compromised because of the higher LOD and requirements of synthesizing new materials (or polymers). Out of these methods, electro-chemical methods are simple, low-cost, more accurate, and easier to field application. Moreover, electro-chemical methods provide a sensitive approach for the determination of many biological and environmental pollutants (Molaakbari et al. 2014; Foroughi et al. 2014; Beitollahi et al. 2014). Thus, recently, electro-chemical techniques have drawn a special attraction in analytical chemistry because of their robust application. But maximum of the electrochemical techniques in melamine determination uses an enzymatic biosensor or molecular imprinting technique (De-Araujo et al. 2014; Wu et al. 2012). Conversely, the overall performance biosensor is influenced by the stability of the enzyme-layer, hence, further calibration of the sensor is always necessary.

Moreover, the thickness of the enzyme-layer can also hider the analytical signals and affects the response-time due to the hindered-transport of the analyte molecules through this enzyme-layer. Thus, these existing techniques are quite expensive and sometimes require highly complicated steps, hence they are time-consuming; even, sometimes poor sensitivity and selectivity have made them inappropriate in the routine determination of melamine (Ai et al. 2009).

We report here a simple electro-chemical approach for melamine determination without using an expensive and complicated instrument based on nanostructure materials modified GCE in PBS 6.0 using two-electrodes system.

METHODOLOGY AND APPROACH

a. Reagents

Antimony(III)chloride, cadmium(II)chloride, ammonium hydroxide, ethanol, disodium phosphate, monosodium phosphate, hydrazine, bisphenol-A, nafion (5% ethanolic solution), acetone, chloroform, ethanol, paracetamol, catechol, 2-nitrophenol, 4-nitrophenol, 4-methoxyphenol, melamine, methanol, and ammonium hydroxide were used as received from the Sigma-Aldrich Company.

b. Apparatus and Instruments

Calcined CAO-NSs was investigated with UV/vis. spectroscopy (Evolution 300 UV/vis. spectrophotometer, Thermo-scientific, USA). The FT-IR spectrum was recorded for the calcined CAO-NSs with a spectrophotometer (NICOLET iS50 FTIR spectrometer, Thermo Scientific, USA). The powder XRD spectrum was recorded with an X-ray diffractometer (XRD, Thermo scientific, ARL X'TRA diffractometer, USA). The morphology of the calcined CAO-NSs was examined by FE-SEM (JEOL, JSM-7600F, Japan). Elemental analysis was explored by EDS from JEOL, Japan. The electrochemical investigations were performed using a potentiostat (Keithley-6517A Electrometer, USA) at room conditions.

c. Synthesis of CAO-NSs by Wet-Chemical Process

CAO-NSs was synthesized by a facile wet chemical process. In a typical reaction process, equimolar (0.1 M) solutions of $Cd(NO_3)_2$, $Sb(NO_3)_3$, and NH_4OH were used. These metal nitrate solutions were mixed in a conical flask for 30 minutes at 50.0^0C with continuous stirring followed by the dropwise addition of 150.0mL of NH_4OH solution (0.1 M) to the mixture. Stirring was continued for another 6 hours at 70.0^0C. On cooling the reaction mixture, a black precipitate of CAO-NSs were obtained. Resulting CAO-NSs were washed with deionized water (DI) and ethanol sequentially and later dried at room temperature for 30 minutes. Then 2 hours of continuous heating at 52.0^oC in an oven produced the as-grown CAO-NSs. Later, the CAO-NSs were calcined at 400.0^oC.

d. Electrochemical Approach

At first, a glassy carbon electrode (GCE) disk with an area of 0.0316 cm^2 was modified with the calcined CAO-NSs using 5% ethanolic Nafion solution and dried it in an oven at 40.0^0C for 2 hours. In a conventional one compartment electrochemical cell (10.0mL), CAO-NSs/GCE and Platinum wire were used as a working electrode (WE) and counter electrode (CE) respectively.

The experiments were performed mainly in 0.1 M phosphate buffer solution (PBS) having pH of 6.0. A potentiostat (Keithley-6517A Electrometer, USA) was employed for all of the electrochemical measurements. Different parameters of the proposed CAO-NSs/GCE based melamine sensor were calculated from the I-V observation.

RESULTS AND DISCUSSIONS

a. Characterization of CAO NSs

The characterization of CAO has been performed by various conventional method such as FE-SEM, EDS, UV/vis., FTIR spectroscopy, XRD, and XPS techniques. Characterization are given in the ESM as (Ω) Optical characterization, (Φ) Structural characterization, (η) Morphological and elemental evaluation, and (ψ) Binding energy evaluation by XPS study in ESM.

APPLICATION: DETERMINATION OF MELAMINE BY CAO-NSs/GCE

a. Selectivity Study and pH Optimization

Figure 1a shows current response of ten environmental pollutants, where (5.0μM; 25.0μL) aq. melamine (red-line) in PBS (pH = 6.0) produce best response by CAO-NSs/GC electrode. How performance of the CAO-NSs/GCE varies with pH values was studied in PBS with different pH values (5.7 to 8.0), which shows that the CAO-NSs exhibits excellent electrochemical performance at a range of pH values. It is obvious that with the change of analyte-solution's acidity, electro catalytic performance of the CAO-NSs with melamine also changes, as it gives variation in current response. In this electrochemical reduction of melamine, pH 6.0 gave the highest current response as in Figure 1(b). Therefore, pH value at 6.0 was selected for rest of the study of this melamine determination by the CAO-NSs/GCE assembly.

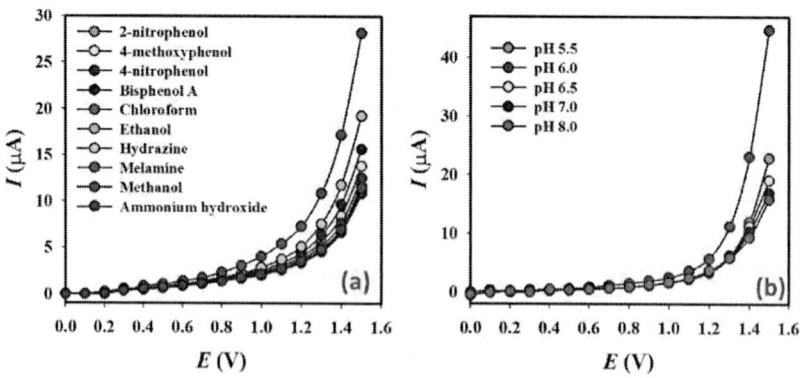

Figure 1. Optimization of the melamine sensor. (a) Selectivity study and (b) pH variation by using 5.0μM melamine solution.

b. Investigation of Melamine Sensor Performance by Electrochemical Method

Figure 2a shows the current responses in PBS (pH = 6.0) obtained from a bare GC electrode (blue-line), and CAO-NSs/GC electrode (red-line).

Current response from the CAO-NSs/GC electrode is roughly 10 fold higher than that from the bare GCE, which proves the exceptional electro-catalytic activity of the aggregated CAO-NSs. Figure 2b shows the current responses of the CAO-NSs/GC electrode in the absence of melamine (blue-line) and in the presence of melamine (red-line; 5.0µM) in 5.0mL PBS. With melamine in PBS, a remarkable decrease in the current response suggests the melamine-sensing ability of the proposed CAO-NSs/GC electrode based sensor. Aqueous melamine of varying concentration (from 0.05nM to 5.0mM) was injected successively to the PBS (5.0mL) and current response variation was recorded after each injection by the CAO-NSs/GC electrode, as in Figure 2c. Gradual decrease of current response with increasing melamine concentration at $25.0^{0}C$ was obtained from the CAO-NSs/GC electrode sensor. Wide range of melamine solution (from 0.05nM to 0.05 M) were taken to define the limit of detection (LOD) and linear dynamic range (LDR) of the proposed sensor. From the calibration plot (Figure 2b), LDR and LOD value was obtained as 0.05nM to 0.5mM ($r^2 = 0.9920$) and 14.0 ± 0.05pM [3 × noise (N) /slope (S)] respectively. While from the slope of the calibration plot, sensitivity was calculated as $3.153\mu A\mu M^{-1}cm^{-2}$.

c. Repeatability, Reproducibility, and Stability

Separately five GC electrodes were modified with the CAO-NSs under identical conditions and employed to determine 5.0µM melamine.

Obtained % RSD (= 3.63%) indicates an excellent repeatability of the CAO-NSs/GCE sensor (Figure 3a). To check the reproducibility, same CAO-NSs/GCE was used in the determination of melamine.

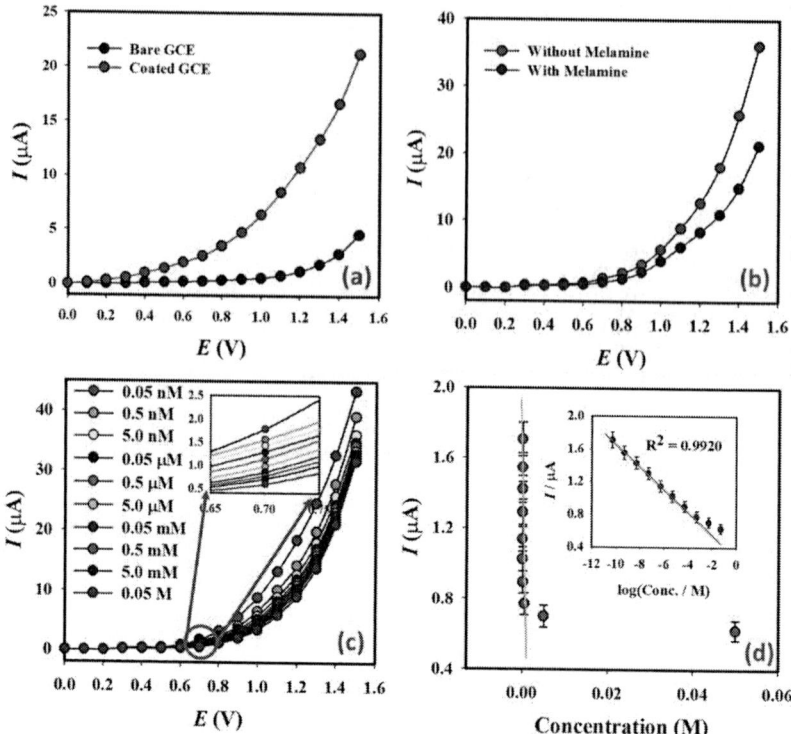

Figure 2. Electrochemical responses recorded in 5.0mL of PBS: (a) Bare GC electrode and CAO-NSs/GC electrode in the presence of melamine (5.0μM), (c) With melamine solution (5.0μM) and without melamine (only in PBS), (c) Concentration variations, and (d) Calibration plot (Inset: current vs. log (conc.).

The almost same result was observed in five repeated experiments with 3.35% (%RSD) for 0.5 nM (Figure 3b) and 4.48% for 0.05mM melamine, which indicates the good reproducibility of the proposed sensor. The CAO-NSs/GCE was stored for 10 days at room temperature. Only 4.13% variation of responses in 5.0μM melamine determination reveals the stability of the sensor.

d. Interference Effect

The possible interfering effect of the common inorganic ions, as well as organic compounds during this melamine quantification, was examined.

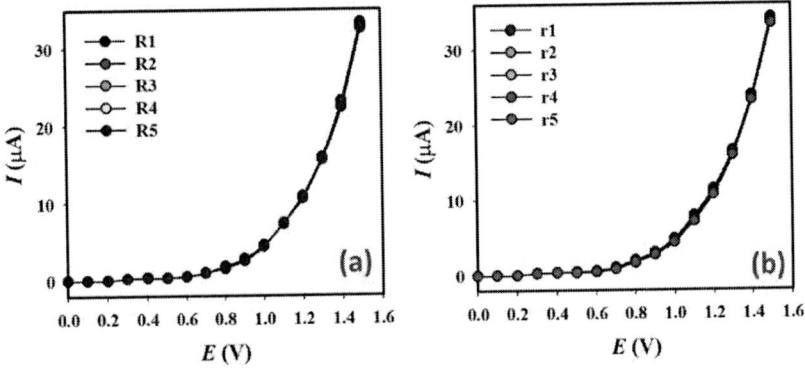

Figure 3. Using 0.5μM melamine in 0.1 M PBS at pH 6.0. (a) Repeatability using five different CAO-NSs/GCE assembly and (b) Reproducibility using the same CAO-NSs/GCE assembly.

The outcome indicates that 400-fold of Na^+, K^+, Ca^{2+}, Mg^{2+}, Pb^{2+}, Fe^{2+}, Fe^{3+}, Cl^-, I^-, NO_3^-, CO_3^{2-}, and 100-fold of 2-nitrophenol, 3-nitro phenol, 4-nitro phenol, 2,4-dinitro phenol, 2-chloro phenol, 3-chloro phenol, thiourea, glucose, creatine, glycine, tyrosine, aspartic acid, lysine, paracetamol, phenol, catechol, 4-aminophenol, 2-aminophenol, 3-aminophenol have negligible interference effect in this melamine (5.0μM) determination (interference $< \pm 5\%$).

e. Real Sample Analysis

To verify the proposed sensor, the CAO-NSs/GC electrode was used in determining melamine in real milk samples (Table 1). For this, we adopt standard addition method to verify the accuracy of the melamine determination in water solution (ten-time diluted milk samples). 25.0μL of aqueous samples of varying concentration and equal amount of real samples were mixed and analyzed in PBS (5.0mL, pH 6.0) by the CAO-NSs/GC electrode as WE. Table 1 represents the outcomes, which indicates that the CAO-NSs/GCE gave a quantitative (~100%) recovery of melamine. So, I-V method is appropriate, reliable, and applicable in real sample analysis with the CAO-NSs/GC electrode assembly.

Table 1. Real sample analyses by CAO-NSs/GCE sensor probe in room conditions

Sample	Added melamine concentration	Determined melamine concentration[a] by CAO-NSs/GCE	Recovery[b] (%)	RSD[c] (%) (n = 3)
Nido milk	0.100 nM	0.1013	101.3	3.4
	0.100 μM	0.1028	102.8	2.7
	0.100 mM	0.1031	103.1	1.9
Baby milk	0.100 nM	0.1017	101.7	3.5
	0.100 μM	0.0973	97.3	2.6
	0.100 mM	0.1035	103.5	2.9

[a]Mean of the three repeated determinations (S/N = 3) with the CAO-NSs/GCE.
[b]Concentration of melamine determined/Concentration of melamine taken.
[c]Relative standard deviation (RSD) value indicates precision among three repeated determinations.

The interior resistance of the CAO-NSs/GCE sensor decreases with the increasing electron-communication, which is a key feature for doped nanomaterial at room temperature and vice-versa (Rahman et al. 2017). In this melamine reduction, On the CAO/GCE surfaces, the melamine molecule is reduced and for this reduction electrons from the conduction band are removed. Therefore, increasing the resistance (decreasing the conduction current) of the CAO/GCE film upon the presence of melamine molecules. Here, in a low melamine concentration, there is a smaller surface coverage of melamine molecules on the CAO/GCE film which leads to a less electron removal from the conduction band and hence the surface reaction proceeds steadily that leads to a higher current response. On increasing the melamine concentration, the surface reaction is increased significantly, which removes more conduction band electrons, hence decreases the conductivity by increasing the interior resistance of the fabricated-film. Because of the flower-like nanostructure, CAO-NSs have larger surface area, which may be responsible for such a sensitive reduction at RTP. The melamine reduction rate in the CAO-NSs was better than other common existing analytes, even under similar conditions (see Figure 1a).

In the electrochemical method, the current response in this melamine determination mainly depends on the sizes and morphology of the nanostructured material. Upon contact with CAO-NSs, melamine gets reduced, which removes the conduction band electrons from the CAO-NSs/GCE. Flower-like morphology of CAO-NSs increases the reduction ability of the CAO-NSs. The CAO-NSs/GCE sensor takes only 10 s to obtain a constant current in this melamine determination. The CAO-NSs/GCE sensor has the higher sensitivity and ultra-low limit of detection than other sensors already published in melamine determination (De-Araujo et al. 2014; Shen et al. 2017; Liu et al. 2015; Chen et al. 2015) as given in Table 2. Due to the large surface area, CAO-NSs presented a positive nano environment in this melamine determination. The proposed CAO-NSs/GC electrode based sensor also produced consistent and stable results. Regardless of these progresses, there are still many key fears that must examine to go for commercial production of the proposed sensor.

Table 2. Comparison of various sensor performances for melamine based on different electrochemical methods

Electrode modification	Methods	LDR/μM	LOD/μM	Ref.
Copper electrode	DPV	5–90	0.85	[2]
Au-Fe$_3$O$_4$ nanocomposites	colorimetry	2–14	-	[3]
Molecularly imprinted electrochemical sensor	SWV	0.4–9.2	0.11	[4]
Gold nanoparticles and reduced graphene oxide	DPV	0.005–0.05	-	[5]
CAO-NSs/Nafion/GCE	I-V	0.05 nM to 0.5 mM	14.0 ± 0.05 pM	This work

Consequently, it can be said that CAO-NSs/GC electrode is highly sensitive in detecting melamine molecules. In this electrochemical measurements, when melamine molecules come in contact with the CAO-NSs, they get reduced by gaining six electrons and six protons as given in Scheme 1. Aqueous melamine was determined by the CAO-NSs modified GCE as a chemical sensor, where melamine gave a significant response. Scheme 1a represents the CAO-NSs/GCE prepared in 5% ethanolic nafion. Theoretical electrochemical response is given in Scheme 1b. Practical electrochemical responses in the presence of the melamine and absence of the melamine onto the CAO-NSs/GCE WE as in Scheme 1c, where a higher current response to the increasing voltage is clearly demonstrated with a delay time of 1.0 s. The probable reduction mechanism onto the CAO-NSs/GCE is given in Scheme 1d; where melamine gets reduced by the removal of conduction band electrons from the CAO-NSs/GCE surface during the electrochemical measurements.

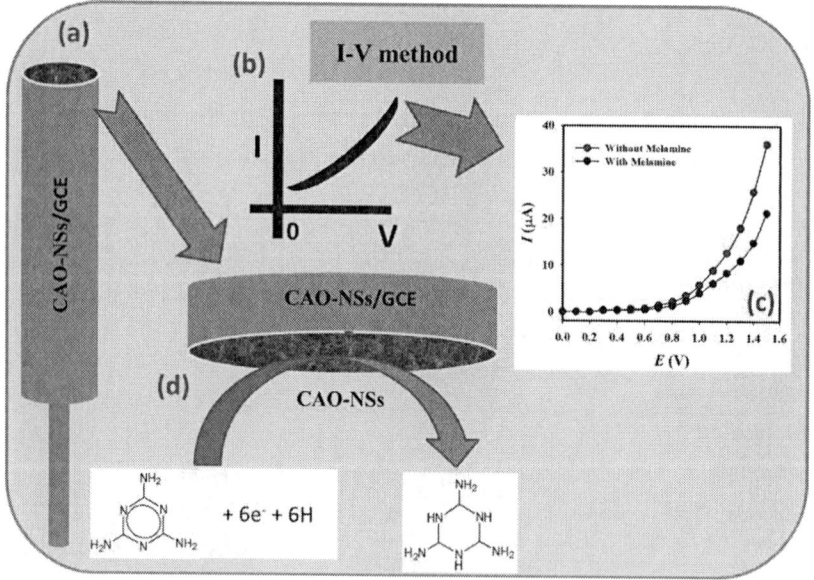

Scheme 1. Scheme representing (a) CAO-NSs coated GCE, (b) Theoretical electrochemical response, (c) Observed electrochemical responses by the CAO-NSs/GCE, and (d) Proposed detection mechanism of melamine, while melamine is reduced by removing conducting electrons from the CAO-NSs/GCE.

Concluding Remarks

Lastly, it has effectively modified GCE electrode by the flower-like CAO-NSs with 5% ethanolic nafion as a conducting-binder to develop a sensitive sensor for melamine determination for the first time. CAO-NSs was synthesized by a facile wet-chemical process at low temperature, which is considered to be the simplest, most convenient, and economical method for the doped-nanomaterial preparation. Crystalline structure, morphological study, optical-properties, band gap and binding energy values were examined by XRD, FESEM, FTIR, UV/vis. spectroscopy, and XPS methods respectively. The proposed CAO-NSs/GCE based melamine sensor was examined by the simple electrochemical method at room temperature. Diagnostic sensor parameters such as sensitivity, LOD, LDR, reproducibility, repeatability etc. were explored thoroughly. Noteworthy research activities including synthesis, characterization and sensing application of the CAO-NSs/GCE towards melamine was included in this report. The proposed CAO-NSs/GCE melamine sensor shows very higher sensitivity and ultra-low LOD with good linearity for a wide range of concentrations in a quick response-time. This noble method could be an active and reliable method of effective chemical sensor development for the determination of various chemicals in health care.

References

Ai, K. L., Liu, Y. L., Lu, L. H. 2009. *J. Am. Chem. Soc.,* 27, 9496 - 9497.
Beitollahi, H., Mostafavi, M. 2014. *Electroanalysis,* 26, 1090 - 1098.
Chan, E. Y. Y., Griffiths, S. M., Chan, C. W. 2008. *Lancet,* 372, 1444 - 1445.
Chen, J. S. 2009. *Chin. Med. J.,* 122, 243 - 244.
Chen, N., Cheng, Y., Li, C., Zhang, C., Zhao, K., Xian, Y. 2015. *Microchim. Acta,* 182, 1967 - 1975.
Choi, J. W., Min, K. M., Hengoju, S., Kim, G. J., Chang, S. I., De-Mello, A. J., Choo, J., Kim, H. Y. 2016. *Biosens. Bioelectrons.,* 80, 182 - 186.

De-Araujo, W. R., Paixão, T. R. L. C. 2014. *Electrochim. Acta,* 117, 379 - 384.

Foroughi, M. M., Beitollahi, H., Tajik, S., Hamzavi, M., Parvan, H. 2014. *Int. J. Electrochem. Sci.,* 9, 2955 - 2965.

Guan, N., Fan, Q. F., Ding, J., Zhao, Y. M., Lu, J. Q., Ai, Y., Xu, G. B., Zhu, S. N., Yao, C., Jiang, L. N., Miao, J., Zhang, H., Zhao, D., Liu, X. Y., Yao, Y. N. 2009. *N. Engl. J. Med.,* 360, 1067 - 1074.

Hirt, R. C., King, F. T., Schmitt, R. G. 1954. *Anal. Chem.,* 26, 1273 - 1274.

Liu, B., Xiao, B., Cui, L., Wang, M. 2015. *Materials Science and Engineering C,* 55, 457 - 461.

Li, X., Feng, S., Hu, Y., Sheng, W., Zhang, Y., Yuan, S., Zeng, H., Wang, S., Lu, X. 2015. *J. Food Sci. C,* 80, 1196 - 1201.

Molaakbari, E., Mostafavi, A., Beitollahi, H., Alizadeh, R. 2014. *Analyst,* 139, 4356 - 4364.

Pan, X. D., Wu, P. G., Yang, D. J., Wang, L. Y., Shen, X. H., Zhu, C. Y. 2013. *Food Control,* 30, 545 - 548.

Pietrzyk, A., Kutner, W., Chitta, R., Zandler, M. E., D'Souza, F., Sannicolo, F., Mussini, P. R. 2009. *Anal. Chem.,* 81, 10061 - 10070.

Rahman, M. M., Alfonso, V. G., Fabregat-Santiago, F., Bisquert, J., Asiri, A. M., Alshehri, A. A., Albar, H. A. 2017. *Microchim. Acta,* 184, 2123 - 2129.

Rovina, K., Siddiquee, S. 2016. *Food Control,* 59, 801 - 808.

Shen, J., Yang, Y., Zhang, Y., Yang, H., Zhou, Z., Yang, S. 2016. *Sens. and Actuator B,* 226, 512 - 517.

Sun, F., Ma, W., Xu, L., Zhu, Y., Liu, L., Peng, C., Wang, L., Kuang, H., Xu, C. 2010. *Trac-Trends in Anal. Chem.,* 29, 1239 - 1249.

Wu, B., Wang, Z., Zhao, D., Lu, X. 2012. *Talanta,* 101, 374 - 381.

Yang, S. P., Ding, J. H., Zheng, J., Hu, B., Li, J. Q., Chen, H. W., Zhou, Z. Q., Qiao, X. L. 2009. *Anal. Chem.,* 81, 2426 - 2436.

Yu, H., Tao, Y., Chen, D., Wang, Y., Liu, Z., Pan, Y., Huang, L., Peng, D., Dal, M., Liu, Z., Yuan, Z. 2010. *Anal. Chim. Acta,* 682, 48 - 58.

Zhang, J., Wu, M., Chen, D., Song, Z. 2011. *J. Food Composition and Analysis,* 24, 1038 - 1042.

Zhu, H., Zhang, S., Li, M., Shao, Y., Zhu, Z. 2010. *Chem. Commun.,* 46, 2259 - 2261.

INDEX

A

acute renal injury, 43
ammelide, 20, 21, 32, 33, 144, 145
ammeline, vii, viii, ix, 2, 4, 6, 7, 8, 11, 15, 27, 29, 30, 33, 58, 131, 143, 144, 145
amniotic fluid, 46, 150
analytical methods, viii, 1, 2, 5, 7, 30, 117, 165
anticorrosive coatings, 81, 86, 87
antimony, viii, xii, 169

B

band gap energy, 125, 126
blood-brain barrier, 150
breastfeeding, 23, 53, 150

C

cadmium doped antimony oxide, viii, xii, 169
calculi, 44, 58, 74, 151, 156
CAO NSs, xii, 169, 173
carbon, xi, xii, 116, 118, 120, 127, 129, 130, 131, 132, 133, 169, 171, 173, 174, 175, 176, 177, 178, 179, 180, 181
central nervous system, 150
chemical, vii, x, 115, 118, 127, 128, 130, 131, 132, 133, 180, 181
chemical sensor, vii, x, 115, 118, 127, 128, 130, 131, 132, 133, 180, 181
composites, 84, 85, 108, 109, 120
counter electrode (CE), 10, 120, 173
crystallinity, 118, 125, 126, 128
crystalluria, 149
cyanuric acid, vii, viii, ix, xi, xii, 2, 4, 6, 7, 10, 11, 15, 23, 28, 29, 30, 31, 32, 33, 35, 37, 44, 45, 46, 47, 49, 53, 58, 59, 63, 64, 65, 68, 69, 70, 71, 72, 74, 75, 131, 141, 142, 143, 144, 145, 146, 147, 148, 149, 151, 152, 153, 155, 156
cyromazine, 3, 7, 8, 29, 31, 33, 34, 35, 43, 69, 71, 73, 74, 146, 159, 162

D

detection, 175
detection limit, xi, 116, 118, 121, 128, 132, 133, 155

dichloroisocyanurates, 146
distribution, xii, 42, 46, 50, 54, 56, 64, 72, 85, 86, 91, 97, 98, 105, 142, 151, 158, 160

E

electrochemical sensor, 117, 118, 120, 127, 128, 179
ELISA, 147, 163
European Union (EU), 12, 29, 60, 66, 116, 170
ex utero contact, 52, 57

F

FDA, xi, 142, 146, 170
flax oil, 81
flaxseed oil, 81
foetal kidneys, 48
foetal skeleton, 49
follow up, 24, 27, 57
food and drug administration (FDA), xi, 2, 4, 9, 11, 12, 30, 31, 33, 34, 59, 61, 66, 142, 146, 148, 157, 170
foodstuffs, v, vii, viii, 1, 2, 4, 5, 14, 28, 37, 58, 66, 75

G

GC-MS, 6, 11, 12, 14, 32, 74, 147, 164
GC-MS-MS, 147
glassy carbon electrode (GCE), xi, xii, 116, 118, 120, 127, 129, 130, 131, 132, 133, 169, 171, 173, 174, 175, 176, 177, 178, 179, 180, 181
growth retardation, 47

H

hematuria, 149
hemiparalysis, 150
hippocampus, 56, 65, 150
human chorionic gonadotrophin, 46
hypochlorous acid, 146

I

in utero contact, 45
infant formulae, ix, 3, 6, 10, 15, 16, 20, 22, 24, 26, 27, 28, 32, 39, 41, 53, 54, 55, 58, 60
infertility, 150
intermolecular hydrogen bonding, 153

K

kidney microtubules, xii, 142, 153
kidneys, ix, 40, 48, 50, 54, 145, 148

L

lactational transfer, 52, 64
LC-MS, 6, 11, 28, 32, 34, 147, 151, 164
LC-MS-MS, 147
LFIA, 147
limit(s) of detection (LOD)
limit(s) of detection (LOD), xiii, 7, 169, 171, 175, 179, 181
linear dynamic range (LDR), 118, 121, 128, 132, 133, 175, 179, 181
linearity, 121, 181
linseed oil, x, 77, 78, 81, 86, 87, 88, 90, 91, 97, 98, 100, 101, 102, 106, 112

M

maternal toxicology, 49

Index

melamin complex formation, 143
melamine, v, vi, vii, viii, ix, x, xi, xii, 1, 2, 3, 4, 5, 6, 7, 8, 9, 10, 11, 12, 13, 14, 15, 16, 18, 19, 22, 23, 24, 26, 27, 28, 29, 30, 31, 32, 33, 34, 35, 36, 37, 39, 40, 41, 42, 43, 44, 45, 46, 47, 48, 49, 50, 52, 53, 54, 55, 56, 57, 58, 59, 60, 61, 62, 63, 64, 65, 66, 67, 68, 69, 70, 71, 72, 73, 74, 75, 77, 78, 79, 81, 82, 84, 86, 88, 89, 90, 97, 98, 106, 107, 108, 109, 110, 111, 112, 113, 114, 115, 116, 117, 118, 121, 127, 128, 129, 130, 131, 132, 133, 141, 142, 143, 144, 145, 146, 147, 148, 149, 150, 151, 152, 153, 154, 155, 156, 157, 158, 159, 160, 161, 162, 163, 169, 170, 171, 172, 173, 174, 175, 176, 177, 178, 179, 180, 181
melamine (1,3,5-triazine-2,4,6-triamine), 170
melamine characterization methods, 143
melamine resin, viii, 2, 42, 60, 146
melamine ware, 60
melamine-cyanurate, viii, xi, xii, 7, 70, 141, 142, 143, 144, 145, 151, 152, 153, 154, 155
melamine-formaldehyde resin(s), vii, x, xi, 60, 72, 77, 78, 85, 106, 142, 146
melamine-related health effects, 143
melamine-urate, 149, 151, 155, 156
melamine-urea-formaldehyde, 81, 108, 111, 112
melaware, 60, 69, 157
methylmelamine, 145
microcapsules, x, 77, 78, 79, 81, 82, 85, 89, 91, 93, 95, 100, 102, 105, 106, 108, 109, 110, 111, 112, 113, 114
microencapsulation, v, x, 77, 78, 81, 82, 84, 85, 87, 89, 101, 102, 103, 108, 109, 110, 111, 113
milk, ix, 2, 5, 6, 7, 10, 12, 14, 16, 17, 18, 19, 20, 21, 22, 23, 24, 25, 26, 27, 28, 29, 30, 31, 32, 33, 34, 36, 37, 41, 44, 45, 52, 53, 54, 58, 59, 63, 64, 65, 67, 71, 72, 74, 116, 150, 156, 161, 162, 163, 170, 177, 178

N

nafion, 118, 120, 127, 133, 172, 173, 179, 180, 181
nanomaterial, 118, 120, 125, 147, 178, 181
nanoparticles, v, vii, x, 87, 102, 103, 111, 113, 115, 130, 161, 163, 179
near-term foetuses, 54
neonatal brain, 56
neonatal kidneys, 48, 55
neonatal liver, 57
neonates, ix, 40, 47, 48, 50, 54, 64, 71, 158
nephrolithiasis, 44, 55, 61, 68, 69, 70, 73, 149
nephropathy, 48, 72, 145
neurological defects, 149

O

oxidation reaction, 131, 132

P

phosphate, 173
phosphate buffer solution (PBS)
phosphate buffer solution (PBS), xii, 169, 171, 173, 174, 175, 176, 177
piloerection, 150
placenta, 45, 46, 50, 54, 71, 146, 149
placental alkaline phosphatase, 46
placental transfer, 45, 68, 159
proteinuria, 149

R

regression co-efficient, 121, 128

relative standard deviation (RSD), 130, 175, 176, 178
reproductive toxicity, xi, 47, 61, 71, 142, 150, 159, 162

S

self-healing, 78, 79, 81, 85, 86, 87, 110, 111, 112
SEM, viii, x, xii, 78, 91, 92, 93, 94, 95, 96, 97, 99, 100, 102, 103, 104, 107, 122, 142, 151, 152, 154, 169, 172, 173
sensor, v, viii, x, xiii, 1, 115, 116, 117, 118, 121, 127, 129, 130, 132, 133, 138, 170, 171, 173, 174, 175, 176, 177, 178, 179, 181
sensor technology, viii, xi, 1, 116, 134
SERS, 147, 155, 156, 160, 164
smart coatings, 78
solution, 173
sources of melamine contamination, 3, 143
s-triazine ring, 143

T

ternary mixed metal oxide, vii, x, 115
thin layer, xi, 116
toxicokinetic study, 46, 50, 54
toxicokinetics, ix, 40, 50, 147
transfer efficiency, 53, 71

U

urea, 108
urea-formaldehyde, x, 78, 79, 111, 113
ureidomelamine, 145
urinary pathology, 149
urolithiasis, 44, 55, 58, 64, 67, 69, 70, 71, 74, 75, 158

W

wash out, 53
wet chemical process, 172
working electrode, 120, 127, 173

X

x-ray diffraction, viii, x, xii, 116, 118, 121, 125, 126, 142, 151
x-ray diffractometer (XRD), viii, x, xii, 116, 118, 125, 142, 169, 172, 173, 181

Z

ZnO/CuO/Co$_3$O$_4$ NPs, vii, x, 115, 118, 119, 120, 121, 122, 123, 124, 125, 126, 127, 129, 130, 131, 132, 133

Related Nova Publications

A Comprehensive Guide to Chemiluminescence

Editor: Luís Pinto da Silva

Series: Chemistry Research and Applications

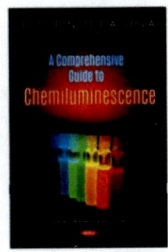

Book Description: This book provides a detailed overview of the basic mechanisms and principles of the most relevant chemiluminescent systems, as well as describing the most recent advances and applications.

Hardcover ISBN: 978-1-53616-170-0
Retail Price: $230

An Introduction to Vanadium: Chemistry, Occurrence and Applications

Editor: Robert Bowell, PhD

Series: Chemistry Research and Applications

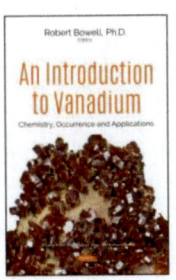

Book Description: Vanadium has, in the last two years, developed new uses, particularly in the field of redox batteries. The expanded use of vanadium reflects the complex redox-related chemistry of the element that allows it to form several oxidation states and to stabilise various chemical compounds.

Hardcover ISBN: 978-1-53616-119-9
Retail Price: $230

To see a complete list of Nova publications, please visit our website at www.novapublishers.com

Related Nova Publications

GAS SEPARATION: TECHNIQUES, APPLICATIONS AND EFFECTS

EDITOR: Suraya Mathews

SERIES: Chemistry Research and Applications

BOOK DESCRIPTION: In recent decades, the science of gas separation by use of a nanoporous permselective membrane has widely developed due to properties such as low energy consumption, easy operation, low waste generation and economic benefits. In *Gas Separation: Techniques, Applications and Effects*, the fundamental concepts of membrane gas separation and the formation of nanoporous membranes are been discussed.

SOFTCOVER ISBN: 978-1-53614-606-6
RETAIL PRICE: $95

CELLULOSE ACETATE: PROPERTIES, USES AND PREPARATION

EDITOR: Calvin Roberson

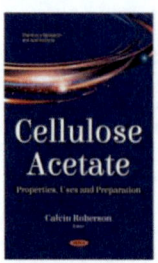

SERIES: Chemistry Research and Applications

BOOK DESCRIPTION: *Cellulose Acetate: Properties, Uses and Preparation* presents data on thermodynamic characteristics (heat capacity, enthalpy, entropy, and Gibbs function) from 4 to 580 K cellulose acetates and cellulose nitrates, as well as the major plasticizers for these polymers, the temperatures of their relaxation and phase transitions, the effect of plasticizers on these characteristics of cellulose acetate and cellulose nitrate and the solubility of plasticizers in polymers.

HARDCOVER ISBN: 978-1-53614-704-9
RETAIL PRICE: $195

To see a complete list of Nova publications, please visit our website at www.novapublishers.com